WHY GOVERNMENT CAN **NEVER** FIX A DOWN ECONOMY

*And Why It Should **Never** Try*

TOM SHIPLEY

outskirtspress

DENVER, COLORADO

Outskirts Press, Inc.
http://www.outskirtspress.com

ISBN: 978-1-4787-1885-7

Outskirts Press and the "OP" logo are trademarks belonging to Outskirts Press, Inc.

PRINTED IN THE UNITED STATES OF AMERICA

Contents

When I began this project, I expected to find the actions of greedy corporations or bankers to have been a major cause of our current economic problems, first announced in December, 2007. When I discovered, to the contrary, that our federal government was the cause, I then wondered how United State's missteps could have upset Europe and the rest of the civilized world.

This book is about the business world in which products and services are made, sold, leased, advertised, etc., and the accompanying financial activities in which money is loaned, borrowed, invested, etc. Together they comprise the U.S. marketplace. These activities are the primary source of all national income – private and public. And public includes city, county, state, and federal government.

It is also about problems that affect these businesses, what has caused depressions and recessions in the past, and what did businesses do to make them spread nationwide. As the facts began to reveal the federal government as the culprit and the reasons became obvious, I introduced them to the tale and explained how the Constitution has allowed those government activities to take place. And finally, we show why we, as a nation, have prospered so well, and how the rest of the civilized world was affected by our activities.

Once I became solidly involved with this project, it was absolutely clear: the history of nationwide business travails in the United States is the story of government's failures in attempts to regulate private businesses and run the economy – total, absolute, costly ignorance.

First, the business world with farmers, manufacturers, restaurants, etc. We will discuss the financial area last. But before we start, consider this. I have been unable to find any instance in which missteps or improper activity by business organizations have caused a problem with our economy. There have been many, many failures of businesses, of course, and many examples of improper activities, but I found nothing that has been done by businesses that affected the entire economy or even a relatively small portion of it. Not even one single example.

The Constitution Was Our Protection

In every instance, the actions of government caused problems that arose and spread nationwide. And in every instance, I found the government to have been operating outside of the limits placed on it by our Founders. And in all cases, government was attempting to solve potential problems that it imagined might occur if it didn't limit business peoples' activities. It was attempting to save some segment of the population from losses due to missteps from risky ventures by businesses or their leaders.

Governments will never understand – businesses' customers are already protected from over-zealous activities by much smarter regulators -- competition and fear of failure. The livelihood of managers and their families is dependent on the health of their firms, and their individual reputations have been built on previous successes. They have no intention of risking their and their families' future on foolish risks. Some will: mistakes will occur, businesses will fail, and managers and their families will suffer. But the suffering will be limited – only that company, its associates, and customers or clients will suffer -- contrary to widespread problems brought on by government regulations and meddling. Government should limit its activities to protecting business customers from the criminal element.

We are suffering from these problems because our federal government -- presidents, congresses, and judges -- over the years violated the Constitution and became active in areas in which they had too little knowledge. In Chapter V we discuss how they did that, and show that those activities were the cause of our ongoing problems today.

The Founders set up a government that was to be ruled by the people, and it gave the government specific areas of operation. To work outside those boundaries, the federal government had to get permission from the people, and a procedure was developed with which to do that. Many think the procedure was developed for difficulty – to make it so difficult to make changes that they would be few in number. But that was not the reason.

The Founders had studied and knew the human failings that had caused the downfall of past governments, and their efforts were directed to avoid those failings in this one. The Founders knew that those in the government would never be the brightest or the most knowledgeable in the country, but the populace would include them in large numbers. They developed a procedure to make sure that those with the greatest minds, inside and outside the federal government, could have a chance to review any Constitutional change the government would like to make, and to make their observations known on the subject. Had the federal government been doing that over all the years, we could possibly have avoided most problems – including our current one.

We have suffered nationwide problems because we – you, me, and our neighbors – stood by quietly while politicians of the worst kind violated the oath they took to uphold the Constitution, and stole the Founder-designed federal government from us.

Government spending has never restarted a down economy; Presidents George W. Bush and Barack Obama were the last to try, and they failed. Many others, worldwide, have tried it and all failed, too. The biggest failures in this country, of course, were Herbert Hoover and Franklin Delano Roosevelt, (1929 to early 1940s). Our presidents have not been alone, however, in their failures; they were assisted at one time or another by famed economists and university professors, members of Congress, and a wide assembly of other acknowledged experts.

Only those directly involved in private-sector business can get a stalled economy moving in a healthy direction. Government can help to make it happen faster, but its activities must be limited: Curtail constraining regulations and let the citizens and private sector keep more of their own money to work with – lower taxes – without time limits. Then get out of the way and stay out of the way. That has been the only process that has been successful during the entire history of this nation – and possibly, (I haven't looked into that), the rest of the world.

Too few of our people – citizens, politicians, economists, educators, etc. -- understand how our economy works. So, they do not understand why we became the most prosperous nation on the face of the earth. In fact, due to the failure of our educational institutions, most of our citizenry don't know that we are and have been for a century or two, the most prosperous. Our Founders gave us the freedoms that allowed us to apply the simplest economic system ever tried; and to conserve them, they set limits on federal governmental activities.

A Natural Economic System

The United States of America's economic system originated in the early 1600s, before the Union was formed. It is, actually, a natural way for people to conduct business and evolved to resolve problems that arose when the system forced upon the Mayflower Pilgrims failed. It was the foundation for the subsequent emergence of this nation as a leader in productivity above all others in the world.

With diligence we have managed to change that system from being recognized as an entirely natural economic system to one known as a capitalist system (economic jargon which would scare anybody), and a free enterprise system, which is more acceptable.

Did you know that in those formative years we tried communism – long before Karl Marx was born? It was a miserable failure. The reasons are revealing, and our children are not being taught that.

To get funding, the contract to which the Pilgrims had to agree, brokered by their merchant-sponsors in London, specified that everything had to be done as a community. The houses the people lived in and the land on which they were built would all belong to the community, and everything, including anything produced on the land, was to go into a common community pool, with each member entitled to one equal, common share. That would eliminate all envy – nobody would have any more than anybody else. Paradise on earth, no less – and President Obama has told us that this is his objective.

On August 1, 1620, the Mayflower set sail for the new world from England. It carried 102 passengers. Forty of them, Separatists led by William Bradford, had left England to escape the oppressive Church of England in search of a home where they could live and worship

God according to their own conscience. The others sought the New World for other reasons. Together, the group became known as the Pilgrims.

Within a short time after arriving and getting settled, the hopes for paradise began to evaporate.

Bradford wrote about this in the style of those days. And, toned down in the Tennessee vernacular, he said that the experience they had tried in several years by godly and sober men, established definitely that the sharing of property and distribution of all individual production into the community for equal sharing did not make the community a happy and flourishing one. It was found to breed confusion and discontent and to retard much industry that would have been of benefit and comfort to the community. The young and most able rebelled at spending so much of their time and strength in working for other men's wives and children without any recompense. The strong men felt that it was an injustice that they received no more in the distribution of food and clothes than that received by the weak who were unable to perform a quarter of the strong man's efforts.

When the problems became acute, the governor, with advice from others of importance, assigned to every family a parcel of land proportioned to the number of individuals in it. And they distributed the children and the weak who had no family connections, to live with those who could accept them, along with their parcels of land. It was ruled that every individual should plant and cultivate as much corn and other products as they wished, and the fruits of their labors were theirs -- and theirs alone -- to dispose of as they pleased.

I say that this has to be the most natural system for transacting business that could be devised by mankind.

Bradford writes of the changes this wrought. "This had very good success, for it made all hands very industrious...." The women went willingly into the fields and took their children with them to plant corn. The community looked at the use of children with different eyes at this point. Previously, to use children in the fields would have seemed a form of tyranny and oppression.

The Pilgrims soon found they had more food than they could eat, so they set up trading posts and exchanged goods with the Indians. The profits they realized allowed them to eventually pay off their debts to the merchants in London. The success and prosperity of this original Plymouth settlement attracted more European settlers, setting off what came to be known as the "Great Puritan Migration."

This early application explains the wonders of the natural way, compared to that of communism or any other. It demonstrated why the theory of communism is flawed and will never work. And contrary to the strife that occurred in the Soviet Union during our time, the religiosity of the Pilgrims allowed them to make the societal changes without bloodshed within about three years. During our lifetime we have seen how communism works in a secular society: When the government leaders see the human weaknesses emerge, they begin to use force to compel people to work better and harder. And eventually, when this doesn't work, the use of force leads to bloodshed.

Note: I first learned of this early history from Rush Limbaugh's book, "See, I Told You So." I checked his facts and found they were correct. In writing this, I used sources from the Internet: www.freerepublic.com; www.libertyhaven.com; www.pngusa.net; and www.pilgrimhall.org.

Capitalism?

Nobody can take charge of, and successfully direct, a nation's economy. Only a free enterprise, free market, economic system has ever really succeeded. Many use the descriptive term *capitalism;* that is just another name for our free-enterprise, free-market, natural economic system. Sometimes, however, those who use the term intend it as an epithet, as though it is something sinful.

I have often wondered about the term *capitalism;* it really does not describe the economic system with which I am familiar. Where did the term come from, and who introduced it? It appears that we don't actually know who introduced the terms capitalist and capitalism. However, I found one answer, and in my search for that, I inadvertently uncovered one of the main tools dishonest, liberal judges have used to change the meaning of our Constitution.

I found the answer in *The Outline of the U.S. Economy,* by the U.S. Dept. of State: http://usa.usembassy.de/etexts/oecon/chap2.htm. "The United States is often described as a 'capitalist' economy, a term coined by 19th-century German economist and social theorist Karl Marx to describe *a system in which a small group of people who control large amounts of money, or capital, make the most important economic decisions.* Marx contrasted capitalist economies to 'socialist' ones, which vest more power in the political system. *Marx and his followers believed that capitalist economies concentrate power in the hands of wealthy business people, who aim mainly to maximize profits; socialist economies, on the other hand, would be more likely to feature greater control by government, which tends to put political aims -- a more equal distribution of society's resources, for instance -- ahead of profits."*

Karl Marx's dream was for the direct opposite of the free-market, natural economic system which has been so successful for us. Our system, with help from our Founders, has been responsible for making us the most prosperous nation on the face of the earth. You can see the results of Karl Marx's dream in the collapse of Russia in 1991, the current economic situation of all countries in the European Union, and our current problems resulting, partially, from our drift toward socialism.

But Marx was successful in one respect: He gave big government, and ignorance, a tool: He is saying business people are bad; they are seeking to acquire wealth (by maximizing profits) so that they can overthrow the government. Think about that. Twenty-nine million firms, including Boeing, GE, Microsoft, Apple, Pratt & Whitney, restaurants, laundries, etc. are all conspiring to acquire power in order to run the government of the United States.

Our current president has used the words "maximize profits" and "equal distribution" several times during a week in June, 2012, as he unfairly described some businesses and people that he blamed for the continuing lackadaisical performance of the economy. Note those words in the Marx quotation. Could our president have been quoting Karl Marx? Some are saying that Obama is very familiar with the works of Marx.

Webster's Unabridged Dictionary, published 1987, provides a current definition of capitalism: *An economic system in which investment in and ownership of the means of production, distribution and exchange of wealth is made and maintained chiefly by individuals or corporations.*

An earlier Webster's Complete Reference Dictionary and Encyclopedia, published in 1942, provides an earlier definition: *The possession of capital, especially its concentration in the hands of a few; the power of combined capital.*

This is an example of the tools that politicians and liberal judges have used in their centuries-old effort to change the meaning of the words in our Constitution. In 1942 our definition of capitalism agreed with that of Karl Marx – dangerous. But by 1987, the definition had changed completely: the danger is gone, capitalism is acceptable. In resolving Constitutional problems to suit their intentions, judges have used the changes in the meaning of words as time went by. And that has caused problems.

Obama was not using an old dictionary, he was quoting Marx. The old definition part was truly in the old dictionary, but "maximize profits" and "equal distribution" were only mentioned in *The Outline of the U.S. Economy* in words attributed to Marx.

Are we smarter than the rest of the world? No. We are *of* the rest of the world; we came from those roots. But we have been privileged, because we haven't suffered the history that other nations have. They were all, at one time, monarchies -- ruled by kings. The king (or a ruler by a different name) directed the activities in his fiefdom. Other countries developed from that, their people endured it and haven't worked too hard to change. We entered the arena much later, with a much different, unique system – devised by our Founding Fathers. They used their experiences and studies of different forms of governments that had been tried throughout world history -- how they had worked and why they failed – to form a different type of government.

Different Government Objectives

The governments of all other nations began with a ruling class – a king or ruler by some other name. And they all earned their kingdoms and their royal blood by conquest. The first king had a better army, was more skilled in warfare, and wanted what the other people had – crops, land, whatever. His immediate family and successors, as long as they maintained a good army, were then rulers. If others, by conquest, took their kingdom from them, the ex-rulers moved away, but retained their royal blood. Their objective was never to make a better government for their people; the kings, their families, and their associates wanted to have what the conquered people had enjoyed. And after conquest, the rulers were primarily interested in maintaining their kingdom and way of life. Sometimes they ruled reasonably and allowed their people freedoms; sometimes they took for themselves all that they could from what the people had.

The record of the earliest kingdoms that I was able to find was Tripods *Ancient Sumer History*.

> ... Sumer was a collection of city-states around the Lower Tigris and Euphrates Rivers in what is now southern Iraq. Each of these cities had individual rulers, although as early as the mid-fourth millennium [4,000] BC, the leader of the dominant city-state was considered to have been the king of the region. ...Although evidence for human presence exists in western Asia far back into paleolithic times [Old Stone Age, 10,000 BC] the *prehistory* of southern Iraq is relatively late in coming; there are no archaeological remains preceding the sixth millennium BC [6,000 BC]....

> ... The Sumerians may have migrated from the East -- either ancient India or Iran -- and were unrelated on the basis of their language to the various groups speaking Semitic languages

in the Ancient Near East... Sumer may very well be the first civilization in the world (although long term settlements at Jericho and Catal Hoyuk predate Sumer and examples of writing from Egypt may predate those from Sumer). From its beginnings as a collection of farming villages before 5000 BC through its conquest by Sargon (Sharrukin) of Agade (Akkad) around 2370 BC and its final collapse from the Amorite invasion around 2000 BC the Sumerians developed a religion and a society which influenced both their neighbors and their conquerors. Sumerian cuneiform -- the earliest written language -- was borrowed by the Old Babylonian Kingdom which also took many of their religious beliefs....

As you can see, this was a little before my time, difficult to understand, but recognizable. This has been the way nations came into existence and faded away. If interested in more information, go to: http://ancientneareast.tripod.com/Sumer.html

The United States of America was the first nation on the face of the earth that was formed with provisions to prevent a ruler-type, kingdom-like system, ruled from the top (the king) down to the citizens. It was designed to be ruled by citizens – all equal in standing, rank, class or status, both citizens and leaders -- designed to be controlled by the people. Since then no other similar government has been formed, and in the current state of affairs in the Sumer neighborhood of the Middle East, we see examples of the age-old methods, just described, of establishing governments and nations: Go in with guns and take charge. Then every leader fights to maintain control of the government. Not one single thought of setting up a government to advance the state and life of their people; all efforts and thoughts are directed to: What can I get for me and my people? No other nation since the ratification of the United States Constitution and the formation of the new government on March 4, 1789, has been planned with the same objectives as that of our Founders: What can we do to make life better for all of the people?

The U.S. Marketplace

THE FOLLOWING DESCRIBES our economic system – simplicity in the extreme – made possible by our Founders. No ruler with a plan; no single director at all. The people of millions of different businesses, each group seeking a way to get the best for its own company, collectively, created the best economy that the world has ever known. And they have been instrumental in improving the livelihood of all other nations.

If the U.S. government would get out of the way now, we could continue that record.

We have approximately 29 million firms in the United States, all competing for their share and more of their market, and their individual activities create our marketplace. Their fortunes, collectively, determine whether our nation is prospering or otherwise. From small neighborhood restaurants, laundries, service stations, barber shops, beauty shops, to local machine shops, mining and construction firms, product manufacturers, to great national, international ones such as General Electric, Boeing Company, etc. All of them have their own capabilities, products, or services to sell, rent, lease, etc., and each of their managers and employees have their own unique expertise. Many, even in the same business, differ completely in skills, marketing area, objectives and procedures.

Companies are Uniquely Different

My wife and I started a business in 1976 -- an advertising business. Using my past engineering experience with automated machines and digital technology, we decided to use it to assist machine manufacturers in their efforts to reach their user customers better and to help them promote their company and product advantages. As technology had evolved from electronic tubes to solid state transistors, to microprocessors, computers, etc., I had seen the need develop; conventional advertising agencies couldn't keep up with rapidly evolving technology. They were trying to sell machine tools, which had definitively useful and different properties, with the same techniques in use for products having no differing features -- such as light bulbs or food products. We decided to fill the niche with an advertising and consulting firm. In 1977, with my engineering experience and my late wife's background in bookkeeping, editing, typing, shorthand, and organization, we began to wend our way, cautiously, toward our objectives.

A friend of mine had an advertising agency, too. In yellow book listings of advertising agencies, it would appear that his business was much the same as mine and my wife's. But they were actually as different as night and day. His background of expertise was merchandising and commercial art, and he intended to use his knowledge of merchandising and his art skills to work with an entirely different set of companies. He, too, proceeded cautiously toward his objectives.

Same general objectives, but different things to sell--completely different markets and approaches. If he had used my plans, he would have quickly failed -- and vice versa. His advertising expertise got him his major client: Vlasic Food Products -- its pickles and food products -- and mine got us business from Hillyer Corp, National Acme, DeVlieg Machine Company, and others -- all machinery manufacturers. However, I often needed his art capability in my work, and he frequently needed my knowledge for help with some industry and technology project.

All Together Now

Each of those 29 million firms, unique capabilities well in hand, charts its own course and aims its people in a selected unique direction. All of those businesses, each using its own capability in thinking, planning, working to satisfy their own individual objectives, have made our economic system outperform all others. The combination of all those millions of individual minds, working on their own unique problems in their own unique way, has yielded results that could never be equaled by directions from one highly intelligent, acute mind or group of minds.

Eventually, we are going to discuss the effectiveness of fear of failure, and how important that is in our marketplace. So, preliminarily, let me cite an example. When my wife and I decided to start our business because of what we perceived as an obvious need, there was another consideration – we, too, had needs. We could not afford to abandon all caution. If I were going to make a big change, we had to make it carefully.

So we made preparations. I continued to discuss the services we could provide to our advertising agency's potential clients. My particular objective was to acquire the annual advertising and promotion business of a machine manufacturer in Mountainside, New Jersey: Hillyer Corporation. I wasn't sure how well they had received my pitch for their business, so I was still discussing employment in machine tool marketing management with various companies – one of them, the Bullard Company. I had rejected its managements' offers twice in the past, but they appeared to be serious for a third time. As time was approaching for my business decision, Hillyer came through. We secured a one year contract for an advertising program. At that point, I was sufficiently sure of the future, so I notified the companies that I was no longer interested in their jobs. The point I'm trying to make is this: The fear of failure kept my wife and me from jumping into the advertising business, even though we saw the need. This is a small, personal example, but in business, the fear of failure is by far the most efficient and effective regulator of managements' activities than anything else that has ever been devised. And as people go to larger companies and make it to management levels, fear of failure is just as much a factor in the way they conduct their completely different operations, as it was with their much smaller ones.

A Single Director for all?

Huge corporations with a wide range of different types of businesses -- General Electric for example – don't fit the mold I have described. (I picked GE because I worked for the firm for fifteen years.) The business description for them, with individual-firm characteristics, does not fit. If the man at the top knows enough to run all those different divisions whose products and markets are entirely different, then some smart dude should be able to run the whole collection of firms that makes up the economy. The answer to that is – the man at the top doesn't. The divisions of those huge corporations are each run like separate companies, and each one has its own managers and people who know that line of business. GE's CEO – the top dog -- has a different, unique set of skills. He and his staff plan a corporate course, with each division's future accounted for. As the months and years unfold, he looks at division reports and numbers. When one of them has missed too many steps in the plans without suitable explanation, he acts. But he doesn't go in and take charge; he selects another manager, skilled in that division's product area to remedy the problems. Sometimes he decides that a division and its product have had their day and sells or otherwise gets rid of it. If the CEO's skills are truly unique, his group of divisions (companies) prosper – and so does he. Not too many business people have these unique skills; good ones are difficult to find. But those who have them are in demand, and they command large salaries.

No central planner can devise a marketing strategy that could be more successful than that resulting from the collective action of the minds of 29,000,000 firms, each of them working to improve his or her own firm's lot. No individual or group can issue advice and develop directions to get those companies, collectively, to a desired economic level which could be defined as best, and none of the 29,000,000 firms are working to make the United States the best. Each of those firms only wants to move its product or service, pay all its workers and other costs, and make a small profit with which to grow. And they, collectively, have made the U.S. economy, by far, the best in the world

Recognized economic experts and journalists wrote widely in the '80s that Japan, with its managed economy, was destined to overtake and surpass the U.S. and its free market economy. It didn't. Japan's economy soon faltered. Professorial economists did not, and still do not, understand what happened. But they have tried to explain it, carefully. And if the work of twelve of them is studied, we will find twelve different answers -- unless some copy the work done by others. If an economist doesn't appear to have most of the answers for economic problems, who needs him? And if the marketplace and how it works is too complex to understand, what is a professor of economics going to teach a student for an advanced economic degree?

Those who have studied business downturns and their causes of the past have noted: Some government misstep, law, regulation, or some combination of them caused the problems. We will present some of government's missteps later on. Missteps by private businesses have caused individual companies to fail, but never has a combination of private business mistakes been big enough, nor have enough businesses individually failed, to cause an economic downturn or panic.

Economists Thomas Sowell and Walter Williams are truly experts, and they could tell this tale more accurately and comprehensively. And there are others of whom I'm unaware who could do the same -- but there won't be many. It is easier to "know it all and tell the world so," than it is to acknowledge that complexity and randomness is too great for precise prediction or explanation.

Hard Times Come

Private sector businesses never see a downturn coming; individual firms only see incoming orders begin to taper off, and at get-togethers with others, it becomes a topic for conversation. If enough other firms begin to see orders taper off, however, and that continues for a time, the nation soon becomes aware: Business quarterly reports, in aggregate, make it evident that overall business is down, and the media lets us know that a possible business downturn or recession exists.

But work on recovery has already started. All of those firms who noted the tapering off of orders or backlogs for a period, have begun to review internal business practices more closely, looking for ways to reduce costs and to improve efficiency, and to dismiss less productive workers. And as it continues, more of those managers whose firms make up our economy will begin to look for ways to stimulate sales or develop new opportunities for their businesses – price reductions, freebies, letters, radio, TV, Internet advertising, whatever. This is all the private sector can do. The next step – a most important one – has to be taken by government.

At this point, a recession is evident. The government can give considerable help, but the only government action that has ever worked to assist the private-sector companies in their efforts has been this: Give the entire private sector a reduction in taxes (reduce tax rates with no time limit) -- for individuals, companies, and corporations -- curtail adverse government regulations, establish a view of the future that is clear, unencumbered by government meddling, and then let all those independent minds, consumers and producers, do their thing. That has been the only process that has worked in the past; only the private sector has ever been able to heal itself.

We sometimes forget; government has no money; it is completely dependent upon the private sector for sustenance, and it doesn't want to give any of that up. So, when it decides to try to bring the private sector back to life, it focuses on one area for growth and uses the tax money from other sectors to help it along. And government, knowing very little about the private sector, will be focusing on the wrong sector. Even business doesn't have the knowledge necessary to pick a winning sector, but it definitely would have known not to go the administration's way.

The Obama administration chose renewable energy on which to foist billions in loans to fix our economy. (Two of his choices of companies soon went bankrupt, and the loans won't be repaid.) He is trying to stop man-made global warming, and nobody has told him that scare has been over for some time. And it was a scam. CO_2 and global temperatures have no relationship to each other, and temperatures from 1850 to 2011 had risen only 1.4 Degrees Fahrenheit. The scare was perpetrated by "scientists" working to improve their way of life with larger government grants, and a popular press that used the scary news to develop interest and readership. Obama also chose to spend the people's money on construction projects in the states, but that failed. He selected projects useful to reelection by going into union territory, and in others, by demanding that bids for the work be increased to union scale, instead of going for the lowest bid. And nothing he has tried has worked. The area in which he spent money was helped, but the rest of the country stayed the same. And when the money stopped coming, it all shut down again.

The Heritage Foundation provides this list of green energy firms the administration selected to provide more jobs and to improve the economy. The money advanced, loaned, or whatever, is now down a rat hole. All of these companies received the sunshine treatment from the Obama administration, and then failed.

Evergreen Solar
Spectra Watt
Solyndra (received $535 million)
Beacon Power (received $43 million)
AES' subsidiary Eastern Energy
Nevada Geothermal (received $98.5 million)
SunPower (received $1.5 billion)
First Solar (received $1.46 billion)
Babcock & Brown (an Australian company which received $178 million)
Ener1 (subsidiary EnerDel received $118.5 million)

Amonix (received 5.9 million)
The National Renewable Energy Lab
Fisker Automotive
Abound Solar (promised $400 million, received 70 million
before bankruptsy)
Chevy Volt (taxpayers basically own GM)
Solar Trust of America
A123 Systems (received $279 million)
Willard & Kelsey Solar Group (received $6 million)
Johnson Controls (received $299 million)
Schneider Electric (received $86 million)
Solar Trust of America
BrightSource (Received 1.6 Billion
Energy Conversion Devices
ECOtality (Received 126.2 Million)
UniSolar
Azure dynamics
Raser Technologies (received 33 Million)
Mountain Plaza, Inc.
Olsen's Crop Service and Olsens Mills Acquisition Co.
Range Fuels
Thompson River Power LLC

Some of the loans went to foreign clean energy companies, but 80% of all loans went to President Obama's campaign donors. For more information:

http://heritageaction.com/2012/07/can-president-obama-name-one-clean-energy-success/

All were favored with loans, selected carefully by our government, because they were in the right business with, the Obama administration imagined, a favorable future. All dropped from view through bankruptcy. As discussed previously, government doesn't have a clue. Only the market picks winners and losers, and only government would try; people in business know every new business entrant to the economy is a crapshoot – regardless of the product or service. People didn't buy enough of these companies' products because they weren't interested in them -- even when government assumed part of their costs. Obama's reaction? "It's Time for Us to Double-Down on Clean Energy That Has Never Been More Promising."

There is no need for "clean" energy, it is too expensive, and there are no advantages to consumers, so they aren't buying it. (A new book, *Man-Made Global Warming? It's Foolishness...,* Tom Shipley, 2011, AuthorHouse, shows "scientists'" claims that humans are causing global warming is a farce.) But Obama happens to know that we are wrong; we don't know what is best for us. He does. (He along with his entire administration are lost in space.)

If government does its job and reduces taxes on all, consumers do their share to help the healing process. Private citizens, using new money they now have because of tax reductions, cautiously listen to business messages, look around, take advantage of some of those freebies and spend a little of the money for products at reduced prices. Cautiously, those who need houses, automobiles, and other major items begin to look around. When their need becomes too great to ignore, a few begin to take the risk to purchase needed houses, or automobiles for transportation. With time, when these customers' purchases have increased sufficiently, the view into the future looks clearer, new orders are brisk, and the backlog is growing--farmers, manufacturers, and services begin procedures to increase output. First, managers keep up by working their people overtime; then, if the backlog keeps growing and they can't handle it with their workers, they start hiring part-time people. Finally, if the backlog of orders continues to grow, they begin to hire new, full time workers. At this point, if enough individual businesses, collectively, are seeing this, the recession is over, unemployment rates have dropped, business growth and government receipts are healthy, and some companies begin to invest in plant expansion and new production equipment.

New hires are considered last, of course, because of cost. It is far more expensive to hire a new worker than it is to pay overtime to an existing employee or to hire a qualified part-time worker. And the only reason any of the procedures are used is because incoming business volume is too great for current employees to handle. When businesses are accused of not hiring in order to "maximize profits," take that as a distinct sign: The accuser is totally illiterate and lost in space.

All do not agree

Many professorial economists do not agree that tax reductions hasten an upturn. With only skills derived from limited business observation, none from actual experience, they have different ideas; they point out that high taxes didn't hamper growth just after World War II and during the Clinton administration in the '90s.

John Trickett, Charleston, Arkansas, recently reminded readers of the Wall Street Journal of the obvious: That we had those good times in spite of high taxes. After WWII the U.S. was the only major nation with manufacturers that were able to function. Europe, the UK, and Asia were rebuilding from war damages, which took years. And in the 1990s, basically all Internet-related technology was centered in the U.S. Victor Davis Hanson, National Review Online, spelled out the details after WWII:

> … The world abroad in 1946 was hardly similar to the world in 2011. …India was a backward colony and in civil turmoil. War-torn China was about to embark on the most self-destructive social experiment in human history. Two-thirds of a centrally planned Soviet Union was in shambles. Western Europe was near starving after years of bombing and Nazi strangulation. The future export powerhouses of Japan and Germany were in ruins. Brazil was pre-modern. The miracles of Hong Kong, Singapore, Taiwan, and South Korea were still imaginary. A victorious Britain was full of self-doubt and exhausted, busy dismantling its colonial empire and nationalizing its steel, transportation, health, and energy industries.

In the immediate postwar years, only a capitalist, self-confident America was poised to supply foreigners with much-needed manufactured goods, expertise, and capital to raise the world from ruin. And from the profits, we were able to pay down our own staggering and unsupportable wartime-incurred debt. Note as well that in 1946 a self-sufficient oil-producing America was not [importing] a half-trillion dollars' worth of [foreign] oil each year.

And J.D. Foster, Heritage Foundation, further explained the situation during the 90s:

"Proponents of tax increases often reference the Clinton 1993 tax increases and the subsequent period of economic growth as evidence that deficit reduction through tax hikes is a pro-growth policy." They say, "President Bill Clinton pushed a major tax increase through Congress in 1993, and … the economy boomed. How, then, can tax increases be so bad for the economy?

"This display of ignorance becomes evident after close scrutiny. Average gross domestic product (GDP) growth was solid (3.2% during 1993-1996), but hardly spectacular for the period. The real economic boom occurred after the 1997 tax cut (4.2% growth during the 1997-2000 period). Job growth was about the same during both periods (11.6 and 11.5 million, respectively). However, wages grew only five cents for the high-tax 1993-1996 period, while they grew 49 cents for the low-tax, 1997-2000 period. Low taxes are still a key to a strong economy.

"What the [tax-increase] proponents do not understand is this: The tax increase occurred at a time when the economy was recovering from recession and strong growth was to be expected…. The evidence is persuasive that the tax increase probably slowed the economy…."

Not only do amateur economists need to know the statistics, they also need to know the situation surrounding them at the time. (For the complete story, go to: www.heritage.org/research/reports/2008/03/tax-cuts-not-the-clinton-tax-hike-produced-the-1990s-boom)

It is truly a display of total ignorance when anybody – regardless of his or her academic degree – thinks out loud that an increase in costs for any business community – large or small -- will improve it economically or instill economic happiness that has been lacking. It is silly to maintain that an increase in taxes, which reduces the money on hand, could cause positive results for businesses. With less money available some activities have to be minimized.

Since our recession hit in 2007, businesses have made the moves and the improvements described previously. Their operations are more efficient, more productive, and earnings are good. But the economy hasn't recovered. The federal government has severely hampered recovery efforts. It introduced too many regulations, many of them to last for decades. It suspended drilling for oil in the Gulf, discouraged drilling in Alaska and all other places under government control; added regulations on the oil industry, the homebuilding, automobile, agriculture and mining industries; tried to stop Boeing from opening a factory in South Carolina; spent extravagantly on a failing green energy gamble; issued new, damaging regulations on electric utilities, and other exercises too numerous to mention. The administration authorized a reduction in social security taxes. The social security tax reduction for workers was obviously temporary, designed as a gift for workers as a reelection ploy, and everyone knows social security revenue is insufficient to meet pension promises. [But an additional point your government has not made clear; the holiday from paying the tax is reducing the amount of future social security receipts citizens will receive when they retire. There is less money going into the pot, so all future retirees will get a little less.] Small businesses, comprising up to 65% of the nation's business jobs, got no help from the reduction; they still had to pay the 6.2% employers' portion as always. Leaving George W.'s tax rates in place was not expected to be a stimulus, but all knew that elimination of them would cause the downturn to worsen.

Harvey Golub gives us the results of government actions: This has been one of the weakest recoveries from a downturn in sixty years.

"… exactly how weak has this recovery been? The Federal Reserve Bank of Minneapolis tracks economic performance for each recovery and compares gross-domestic-product growth and job growth, the two most important indicators of economic performance. Over the past sixty years, there have been eleven recessions and eleven recoveries.

"Sadly, this recovery is near the bottom of all eleven. Cumulative nonfarm job growth is just 1.9%, 34 months into recovery, the ninth-worst performance and well below the average job growth of 6.5%. Cumulative GDP growth is just 6.8%, eleven quarters into this recovery, less than half the average (15.2%) and the worst of all eleven.

"But wouldn't things be even worse without massive fiscal and monetary stimulus? It's true that monetary policy by the Federal Reserve has resulted in extraordinarily low interest rates, almost zero for the past three years. Normally, low interest rates would result in increased borrowing by individuals and businesses, generating increased economic activity. Its positive effects in this recovery, however, have mainly been to help the government borrow more cheaply, large banks recapitalize quickly, and homeowners refinance at low rates.

"Uncertainty regarding ObamaCare and higher taxes on businesses and individuals has discouraged the type of borrowing and lending that low rates generally encourage. Near-zero interest rates have also resulted in historically low yields on savings and encouraged riskier investments. In effect, we have subsidized increased spending by penalizing savings…."

(These are excerpts from Harvey Golub's article, *How the Recovery Went Wrong*; May 23, 2012, The Wall Street Journal.)

A solution that would stimulate recovery of the economy has been proposed. Revise our tax structure; eliminate the tax breaks given to many corporations, business friends, favored sectors of the economy, and others. All of us are familiar with the favors given to them because of the business they are in and the government people they know or have known, and this would allow us to reduce the tax rate from 35% to 19 or 20%, yet still yield more income from corporations. Currently, most small companies, which comprise 65% of our business community, are paying the rates that individuals pay. Studies have shown that on day-to-day purchases that we make, the prices we pay include the costs for business taxes and regulations, and the costs are high. The prices we pay for groceries, kitchen accessories, refrigerators, books, etc. have been increased, one study says, by 22%, another says by 23%, to pay for the costs of regulations and taxes.

If we reduced the tax rate to 19 or 20 percent, stopped the favoritism to some and the war on others, removed unreasonable regulations, and gave a small reduction in individual tax rates, we would quickly see a turnaround.

Absolutely nothing has been done since 2008 that could be expected to improve the business situation. American firms can see no clear road ahead; they see only a meddling federal government that has its own objectives and agenda. And the excessive spending, obvious to all, seemingly cannot be stopped – in the U.S. or Europe

A More Definitive Program for Recovery

Jeffrey Miron, senior fellow at Cato Institute, has studied the problems and has a reasonable procedure which history has shown to be the only type of program that has ever eased us out of a downturn. Mr. Miron says:

It is a program that will restore U.S. economic performance:

- Cancel all the tax increases scheduled to take effect at the end of 2012 and provide tax stability going forward. Make (all) the Bush tax cuts permanent. Repeal the alternative minimum tax. Eliminate the health care law's increases in the hospital insurance tax. All this will stimulate in the short term and set the stage for long-term growth.
- Reform the tax code by eliminating the misguided deductions, exemptions, credits and loopholes that distort incentives and

reward special interests. These features include big-ticket items like the deductibility of mortgage interest and employer-paid health insurance premiums, plus myriad small but senseless other provisions.

- Lower the corporate income tax rate. The U.S. corporate tax environment is one of the least business-friendly in the world. Driving investment overseas cannot be good policy.

- Slow the growth of entitlements. The U.S. can afford a social safety net, but our current programs are not sustainable, even in a robustly growing economy. Everyone should agree, at a minimum, to cuts that are sufficient to prevent these programs — Medicaid, Medicare, and Social Security — from bankrupting the country.

- Embrace immigration. Despite recent difficulties, the United States is still an attractive destination for those seeking a better life. By expanding immigration for low-skill workers, we restrain labor costs and reduce out-migration of manufacturing and other business. By easing immigration for high-skill workers — many of them trained in the United States at taxpayer expense — we get a return on our investment and retain industrious and innovative people.

- Scale back military involvements around the world. A strong national defense makes sense, but it must focus on protecting the United States, not paying for Europe's defense or trying to force democracy down the throats of countries that are not receptive.

- Cease the campaign against carbon-based fuels. Green energy may have its day, but only when coal, oil, and gas become truly scarce. In the foreseeable future, traditional energy is much cheaper, and subsidies for alternative energy are a waste.

- Stop scapegoating the rich and pretending that tax-hikes on the 1% can balance the budget. Everyone knows the numbers do not add up.

- Respect capitalism. Anti-business rhetoric, which casts all success as undeserved, and which fails to recognize the improvements in material well-being that result from entrepreneurial success, just drive away talented people and guarantee our economic demise.

  If the United States adopts these policies, it will not only attain [a] 6% unemployment goal, it will once again be the economic beacon of the world.

And he says: "What's wrong with that?"

Read it all at: http://www.cato.org/publications/commentary/how-get-economy-growing-fast

Corporations…Are They Really Bad?

From recent media reports, it appears that many of our citizens have lost an understanding of what a corporation really is.

Corporations. Greedy, greedy corporations. They are continually hiring lobbyists in efforts to pressure our good government people to give them favorable treatment. And as Marx has advised us, they are in league to take over the government of the United States. Fortunately, Washington politicians tell us, they resist their greedy efforts; without our friends in Washington, corporations would overrun and ruin the country.

What did they do to get that reputation? Corporations are the source of approximately 35 percent of the total jobs for our people in this country, and I estimate their employees are the source of about 35% of the individual income taxes which politicians use to pay the nation's bills and politicians' salaries. Also, in 2011 corporations' taxes contributed 5% for the nation's expenditures for the year. Individual income taxes paid for 30%, Social Income taxes, not intended for that purpose, paid for 23% of the expenditures. Other taxes -- excise, estate and gift taxes, customs duties and other miscellaneous receipts paid 6%. The total income from all these sources was $2,302, 000,000,000, and was 64% of all expenditures. We borrowed $1,299,000,000,0000 -- 36% of our expenditures – in order to pay all bills.

It appears to me that businesses have kept us running – they do some good -- and it appears, from the borrowing problem, as though we need some more corporate help, as well as to curb spending.

But are they really as bad as some politicians would lead us to believe? Let us take a look.

A man for whom I have much respect is Jack Welch. He ran the company for which I worked, but not while I was there; I left in 1965, he arrived in 1981. He and his wife, Suzy, wrote an article explaining what corporations are, for the Wall Street Journal, July 16, 2012 issue. This is an excerpt:

> … Elizabeth Warren introduced President Obama at a big fundraiser in Boston: "Mitt Romney tells us, in his own words, he believes corporations are people. No, Mitt, corporations are NOT people," she pronounced. "People have hearts. They have kids. They get jobs. They get sick. They love and they cry and they dance. They live and they die. Learn the difference." The audience went wild.

> What nonsense.

> Of course corporations are people. What else would they be? Buildings don't hire people. Buildings don't design cars that run on electricity or discover DNA-based drug therapies that target cancer cells in ways our parents could never imagine.

> Buildings don't show up at a customer's factory and say, "We won't leave until we solve your inventory problem." Buildings don't encourage their employees to mentor inner-city kids in math and science. Buildings don't fund homeless shelters in Boston or health clinics in Rwanda. People do.

> Corporations are people working together toward a shared goal, just as hospitals, schools, farms, restaurants, ballparks and museums are. Yes, the people who invest in, manage and

work for corporations are there to make a profit. And yes, corporations may employ some bureaucrats, jerks, cheapskates and even nefarious criminals.

But most individuals working in corporations are regular people, people just like you and your friends and neighbors. People who want to make a living and want to make a difference....

A corporation can be viewed as a village consisting of a large number of people whose interests and activities are directed toward one goal – to advance the state of their lives by, collectively, producing a product or products in which others would be interested. Corporations have no interest in running the country, contrary to some politicians' claims. When a corporation's management seeks something from the government, it does so in an effort to advance the interests of all of its employees – management included -- and the public will never find a more comprehensive political cross-section – Democrats, Republicans, Independents, etc. -- of American citizens elsewhere. However, they do use lobbyists; government makes it necessary. Mostly, earlier on, they used lobbyists in defense – to prevent government from hampering their production or sales efforts with regulations or taxes, or improving their competitors' efforts by giving them (competitors) favors that could harm the future for the corporation, its products, and its employees.

Microsoft learned. Government had nothing for them, and in its early days, it didn't contribute for several years. But two companies, competitors who were not faring so well, got with a Republican senator. Time passed, and Microsoft now contributes well. Newly arriving corporations learn quickly – it is contribute or suffer from government activities.

As government got bigger, and began to hand out greater and greater portions of their pie, corporations began to line up to get theirs. The pie was sweet, but as some great man once said, if government can give you something, it can also take it away. Corporations displayed their aptitude for pie with Obama's healthcare plan; some were told of wonderful things that were in the plan for them. They went for it in a big way -- that was before the reading. The health insurance and drug companies now know its contents, and are now learning the government lessons of which the Founders warned us.

The Supreme Court recently spoke, revoking a 2002 law that banned corporate spending in elections. The 2002 law, a Republican/ Democrat creation, usually called McCain-Feingold, banned the broadcast, cable or satellite transmission of "electioneering communications" paid for by corporations or labor unions from their general funds in the thirty days before a presidential primary and in the sixty days before general elections.

The Court ruled that a corporation has the same freedom of expression as that held by an American citizen. The corporation can speak, just as any citizen, with its money – within limits, of course.

Some members of Congress and the administrations do not favor the freedoms that businesses have had since 1789. They do not understand that a corporation is composed of a wide assortment of workers' with widely varying political views and that it is generally non-partisan in nature. Questions that are never asked of an applicant for employment: Are you a Democrat or a Republican? All questions asked involve background, education, past experience – anything to indicate his or her potential as a good worker. It becomes strongly partisan, however, when government actions threaten the corporation's welfare, or when a corporation needs favors from the government (think about the needs of those involved in sun and wind power). Many in our government do not believe the capitalist system, our free economy, which made United States citizens the most prosperous of the world, is fair. Reason? Some citizens have not prospered, while others have prospered too well.

Unless you read the Wall Street Journal, National Review, observe Fox News, listen to talk radio (like WJR, Detroit), or patronize the Internet, you don't get sufficient national and international news, and have no way to realize how blessed we have been and are. The Wall Street Journal, in 2004, reported it, and these are excerpts.

[A] study, EU vs. USA, was done by a pair of [Swedish] economists--Fredrik Bergstrom and Robert Gidehag--for the Swedish think tank, The Timbro. It found that if Europe were part of the U.S., only tiny Luxembourg could rival the richest of the fifty American states in gross domestic product per capita....

...Higher GDP per capita allows the average American to spend about $9,700 more on consumption every year than the average European. So Yanks have by far more cars, TVs, computers and other modern goods. "Most Americans have a standard of living which the majority of Europeans will never come anywhere near," the Swedish study says.

But what about equality? Well, the percentage of Americans living below the poverty line has dropped to 12% from 22% since 1959. In 1999, 25% of American households were considered "low income," meaning they had an annual income of less than $25,000. If Sweden--the very model of a modern welfare state--were judged by the same standard, about 40% of its households would be considered low-income. In other words poverty is relative, and in the U.S. a large 45.9% of the "poor" own their homes, 72.8% have a car and almost 77% have air conditioning--which remains a luxury in most of Western Europe. The average living space for poor American households is 1,200 square feet. In Europe, the average space for all households, not just the poor, is 1,000 square feet.

U.S. citizens' scale of living is the envy of the world, and the Swedish economists have explained why. (This information has been available from other sources, but the two Swedish economists are beyond accusations of slanted results in the United States' favor.) Our average poor citizens' living conditions are on an equitable basis with that of the average of citizens, rich and poor, in the European Union. And those members of Congress, some members of the Court and administration, obviously in the dark also, are diligently, and ignorantly, trying to make us and our economy look just like that of the EU.

The private sector, corporations and small businesses, and individual citizens have achieved this with the help that our forebears left us – freedom. Government efforts should be directed toward preserving our freedom and economy, not changing it for something else that history has shown to be far inferior.

The Swedish study, in entirety, is available on the Internet: http://www.timbro.se/bokhandel/pdf/9175665646.pdf

I had information on this country's superiority before the Swedish guys entered the picture. You may not need it, but I have to give it to you, anyhow.

Things I Didn't Know in 1944

About ten years ago, I found an old book I had inherited from my parents' in 1994: *America Unlimited,* by Eric Johnson, published in 1944, while World War II was going on. The author presents the state of our country preceding and during the war years—and I was surprised at what I read. He describes our successes and explains the reasons for them. And, further, explains why England, Germany, Italy, France, etc., had lagged so far behind us, and would lag behind in the future.

I didn't know: Before the war began, the United States, with less than 7 % of the world's population, "…used 35% of the world's railroads, 45% of all the world's radio sets, 50% of all the world's telephones, [and] 70% of all the automobiles. …It consumed 56% of all the silk in the world, 59 % of all the petroleum, 50% of all the rubber, 53% of all the coffee, and 21% of all the sugar."

"A U.S. Department of Agriculture report showed that 85 out of 100 farm families owned a motor vehicle. In France and in England, before the war, there was one automobile to 25 persons; in Germany, one car to 55 persons; in Italy, one to 109; and in most of the rest of Europe, Asia, Latin America, the proportion ... [ran from] one to 1000 to 10,000 [persons]. In America, one of every five persons had an automobile."

It came as a shock to me that in 1939 we were more prosperous than the people of the rest of the world. I thought of Kingsport, Tennessee, native Ben Sullivan's September, 2003, comment, made during an interview for an article for the high school's newsletter, concerning conditions in those days, "Nobody had any money [back then]—nobody." And that truly told the story – yet we were better off than the rest of the world.

Johnson goes on. "Nearly half the total manufactured goods in the civilized world are produced in the United States. Studies made by the League of Nations in 1929, when the whole world was at a maximum level of production, showed that the income of the United States was as large as the income of the next seven or eight richest countries combined—among these [were] Great Britain, Germany, Russia, France, and Japan."

Today, we are reminded daily by the liberal element that we are consuming more than our fair share of the world's goods and should be ashamed of ourselves. But I'm not; we can do so because we produce more than the rest of the world. And our immigration picture says that many of the people in the rest of the world want to join us.

When the war started, "We mobilized and equipped ten million men for war, equipping them with the most advanced modern weapons, in the space of less than two years. The time element needs to be underscored. Germany...had channeled everything into military preparations for nearly seven years before it launched World War II; Russia had deprived itself of necessities, not to mention comforts, for nearly twenty years, so that it might divert all its energies and resources to military purposes. America, so far as the implements of war are concerned, started from scratch on the day when a program of Lend-Lease [March, 1941] was decided upon.

"Having withdrawn ten million men from our productive process, we [were] able, notwithstanding, to gather and train the man power needed to equip and feed not only our own forces, but to help feed and equip all our allies. I shall not burden these pages with ...the statistics of American Lend-Lease contributions to Great Britain, Russia, China, even Latin-American nations. [But] on every battlefield of the globe, American products made up margins of shortage and tipped the scales of victory.

"...Yet enough margin of industrial capacity remained to keep civilian standards of living and consumption incomparably higher than [that] in any other belligerent country and higher than most of the experts had thought possible. The aggregate of goods and services available to civilian consumers in 1943, indeed, was greater in value than in the first war year. ...In a broad sense, we have [had] both 'bullets and butter.'"

The author, Eric Johnson, born in 1895, came from a poor family. He served in the Marines during and after World War I. In 1921 he was injured and had to retire from the military. The doctors told him to get outdoor work, so he sold vacuum cleaners. Before long he began to repair them, and got into the electrical field. He started a retail shop for cleaners and washing machines and two years later purchased the largest electrical contracting and retail appliance firm in the Northwest. In 1929 he headed a trade organization and in 1931 became president of the Spokane Chamber of Commerce. At the request of a bank, he undertook to salvage the Washington Brick, Lime, and Sewer Pipe Company which, in 1933, a year of gloomiest depression, was in debt for $260,000. Ten years later, as he was publishing this book, he returned the company to its owners completely in the black, with money in the bank and production in full blast. During that time, its minimum wage had risen from 37.5 to 86 cents an hour. In 1942, he assumed the presidency of the U.S. Chamber of Commerce and there, dealt with business and labor leaders, Congress, and President Roosevelt.

I mention these things about his career in order to show that he wasn't a light weight. He says, "...The most revealing fact about that

penniless, work filled, boyhood of mine is that it was devoid of re-
sentment at the time and left no sediment of bitterness afterward.
It did not occur to me that I was the victim of society; that there
was anything the least bit shameful about being poor; [that] the folks
who exchanged their pennies for my papers were 'exploiting' me.
Hardship did not stir me to revolt; it only served as a spur to ambition.
Not once did it cross my mind that I might remain forever stuck in the
mud of poverty."

He says, "What distinguishes the American system from all others
is that it recognizes the individual as the pivot of society. Not classes
as in Britain and other semi-aristocratic societies, not the aggregate
population as in Germany and other totalitarian societies, but the
individual man, woman, and child. Those who fail to understand this
unique American feature prattle of greed and obsessions with comfort
as the moving forces in our life. They do not grasp that individualism
is a national philosophy mirrored in our laws, our institutions, our
mores and prejudices."

He quotes an Englishman, James Fullarton Muirhead, writing
about America for [the benefit] of his countrymen in 1900. Muirhead
wrote that Americans have "an almost childlike confidence in human
ability and fearlessness of both the present and the future, a wider
realization of human brotherhood than has yet existed; a greater the-
oretical willingness to judge by the individual than by the class, a
breezy indifference to authority and a predilection for innovation, a
marked alertness of mind and a manifold variety of interest, above all,
an inextinguishable hopefulness and courage. It is easy to lay one's
finger in America upon almost every one of the great defects of civilization—even those defects which are especially characteristic of the
Old World…In a word, America has not attained, or nearly attained,
perfection. But below and behind and beyond all its weakness and
evils, there is the grand fact of a noble national theory, founded on
reason and conscience."

This book is full of good stuff, and I hope you get the feeling,
from these fragments, of its power. But I must write about one more

thing—the competition that Johnson envisioned [after the war was over] between economic and political systems. And, remember, this was written in 1943/44.

"I am not alarmed by the existence of other economic and political systems and have no fanatic urge to make this one uniform world. ...Russia and the United States represent two extremes, and their inevitable competition for world markets will have about it something truly titanic. One is history's greatest democratic capitalist society, the other the greatest collectivist society ever known; one is people's capitalism at its best, the other state capitalism at its strongest. Americans do not shrink from that mighty test.... Fully aware of Russia's emerging force and productivity, we confidently put our faith in the American way of life."

We now know that Russia, later the U.S.S.R., the power of state capitalism (communism), lost that economic struggle forty-seven years after this book was written. And for reasons foreseen in the book. But there was one other bit of history that intrigued me.

"During the historical three-power WWII conference [United States, Great Britain, and Russia, in December, 1943] in Teheran, [the first conference attended by Stalin], Marshal Joseph Stalin found occasion to remark that without American production the United Nations could not win the war. It was the statement of an obvious fact, but dramatic in that setting and from that source. ...A few months later the head of the American communist movement, Earl Browder, officially postponed 'the revolution' in the United States of America. He announced his party's sad conviction that free enterprise remains the dominant force in this country; he even advised his embattled legions to 'lay that pistol down' and make truce with the capitalist order...."

It was revealing to review Eric Johnson's explanations, expressed almost seventy years ago, of the reasons for our successes. He says they resulted from our way of life and the optimism bred from the freedom that our capitalistic economy and political system allow. We know, sixty-eight years later, that these factors have enabled all of us to advance markedly since his time.

His story, told so long ago, was a welcome change from the current criticisms of our country that imply our successes have been built on the backs of others, using improper methods built on greed. And Eric Johnson warned that these criticisms, arising from ignorance of the facts or from envy, would grow—in this country and elsewhere. But the fact is, even today, we enjoy this highest standard of living because we have been and still are the most productive nation on the earth.

Note: this 1944 information, in its entirety, was developed from the contents of the book *America Unlimited*, by Eric Johnson, published in 1944 by Doubleday Doran and Company, Inc. The book is out of print, but used copies are available from www.amazon.com; search for Eric Johnson, America Unlimited.

Our President has been complaining about the enormous profits that the oil industry corporations are enjoying during these bad times.

Contrary to some foolishness spewed by politicians, I have never met anybody in small or large business whose objective was to "maximize profit." I had never heard the term until President Obama made his first speech. The objective of every manager and worker of every manufacturing business that I have ever known or have been associated with was to move the product. Prices were set and expenses determined and six to seven percent after taxes was generally the goal. And when the popular press screams about indecent profits and tells you how many millions or billions it is, notice the absence of sales magnitude or percentage of profit to sales. They are always missing, and I have wondered: do the reporters not understand the math, or is this their way of creating interest? I have noticed, however, the Wall Street Journal always provides the information.

It is possible the Senate Democrats don't understand the math either:

The top five oil companies (Chevron (<u>CVX</u>), Shell (<u>RDS.A</u>), <u>BP</u>, ExxonMobil (<u>XOM</u>) and ConocoPhillips (<u>COP</u>)) are part of the "Major Integrated Oil and Gas" industry, and the CEOs of the "Big Five" are appearing today [May 12, 2011] before the Senate Committee on

Finance, to get grilled about the "taxpayer subsidies" and "tax breaks" they receive, explain why they deserve to earn record "windfall" profits, and explain the role they play in higher oil and gas prices.

As the table below shows, the Integrated Oil and Gas industry made an average profit of 6.2% -- that is 6.2 cents per dollar of sales, which ranks #114 out of 215 industries by profit margin, and puts oil companies right in the middle of industries by profitability.

The following table shows some of the industries with higher profit margins than the oil industry

Rank	Industry	Profit Margin
101	Confectioners	7
101	Aerospace/Defense - Major Diversified	7
102	Publishing - Newspapers	6.9
103	Jewelry Stores	6.9
104	Home Health Care	6.9
105	Computer Based Systems	6.9
106	Pollution & Treatment Controls	6.8
107	Lodging	6.8
108	Communication Equipment	6.7
109	Aerospace/Defense Products & Services	6.6
110	Sporting Activities	6.5
111	Packaging & Containers	6.5
112	Catalog & Mail Order Houses	6.5
113	Drugs - Generic	6.3
114	**Major Integrated Oil and Gas**	**6.2**

(http://mjperry.blogspot.com/2011/05/oil-profit-margin-ranks-114-out-215.html)

These Senate politicians, with the Obama Administration, are supposed to be developing business regulations to make life better. These five oil companies are among the twenty-nine million firms that comprise our free market. You can see that their progress toward a good economy is being hampered by the ignorance and activities of government. You can also see why this type of activity has stifled the efforts of the private sector in its attempts to improve conditions, and why the downturn continues.

The government takes more in taxes from the oil industry than the industry sees in profit. The *Wall Street Journal*, Thursday, March 15, 2012, brought light to the subject.

Obama says, "They are not paying their "fair share". Here's a staggering fact: The Tax Foundation estimates that, between 1981 and 2008, oil and gas companies sent more dollars to Washington and the state capitols than they earned in profits for shareholders.

Exxon Mobil, the world's largest oil and gas company, says that in the five years prior to 2010 it paid about $59 billion in total U.S. taxes, while it earned . . . $40.5 billion domestically. Another way of putting it is that for every dollar of net U.S. profits between 2006 and 2010, the company incurred $1.45 in taxes. Exxon's 2010 tax bill was three times larger than its domestic profits. The company can stay in business because it operates globally and earned a total net income after tax of $30.5 billion in 2010 on revenues of $370.1 billion.

And this does not include the additional taxes the federal government and states collect at the gas pump. Somebody better suggest to the president that he needs to go back to about the 4th grade in school, and finish his education.

Fear of Failure

In January, 2009, the government acted once again to remove the only regulator of business practices that has ever been found. It stepped in with the peoples' money to save failing companies. Politicians don't know it, Democrat or Republican, but with that action, government removed the only successful business regulator that the world has ever known: The fear of failure for large, too-big-to-fail corporations. The successful managers of each of those millions of private companies have always curbed ambitions; in trying to keep their companies viable, they didn't dare to take huge risks. They operated carefully because bankruptcy and liquidation would be a terrible occasion. They didn't want to fail. And that uniform fear, in the vast majority of all companies that make up the economy, had been the major regulator that had worked for a good economy. The government has now given the larger ones the incentive to go for broke.

Brown Brothers Harriman, Wall Street's oldest surviving general partnership, shows another, better way; it expands the fear of failure concept and gets public money and politicians out of the regulation picture. We cover that later.

When Competition Is Lacking

General Motors and Chrysler, the administration said, were too big to fail. However, absolutely no proof of this theory has been offered, or was even called for by the popular media. So the government did its thing. Now, with governments' supervision, is it possible they can't fail?

Let's look at the underlying cause of GM's, Chrysler's, and Ford's problems, which had existed for some time. Let's look at the entire automotive picture. In the '60s, '70s three companies, and only three, reigned in the U.S. automobile market -- Ford, GM, and Chrysler. They competed with each other. And the United Auto Workers (UAW) union served them all. At negotiation time, the union first selected the weakest company to concentrate on, and then pedaled the decisions to the other companies. But eventually, it took the companies, one at a time, in order – if it was Ford's time, and the UAW wanted it to dance a jig – no problem. All knew that GM and Chrysler would dance the same jig that the union had orchestrated with Ford. Nobody wanted a strike, and all they needed to do was agree to the union's demands and then all raise the prices on their products to pay for the new benefits. So, all three raised their prices and danced to the particular jig which had been acceptable to the first.

For years the Union got its way. (Do you remember when you were advised not to buy a car assembled on Monday? Did you know that that advice was good, and it resulted from the fact that too many of the auto workers partied too heavily during the weekend?) The companies were powerless; they couldn't weed them out as bad employees; bad guys couldn't be fired. Only the union decided who was bad. But no worry, all three companies in the auto business market had the same problem and could live with it.

What a joyous place to work. If you got laid off, for lack of work – no problem. Go sit down in a place made for you and hundreds of others, and rest. You didn't get 100% of your pay, but it was enough to keep you and your union relatively happy.

Then came the foreigners – invited in by the federal government. And they took over. Those union rules which they had accepted for decades brought them down, and Obama's fifty-four mile per gallon stipulation clouded the future even further.

President Obama announced on July 29, 2011, a historic agreement with thirteen major automakers to pursue the next phase in the Administration's national vehicle program, increasing fuel economy to 54.5 miles per gallon for cars and light-duty trucks by Model Year 2025.

These programs, combined with the model year 2011 light truck standard, represent the first meaningful update to fuel efficiency standards in three decades and span Model Years 2011 to 2025.

Together, they will save American families $1.7 trillion dollars in fuel costs, and by 2025 result in an average fuel savings of over $8,000 per vehicle. Additionally, these programs will dramatically cut the oil we consume, saving a total of 12 billion barrels of oil, and by 2025 reduce oil consumption by 2.2 million barrels a day – as much as half of the oil we import from OPEC every day.

How does President Obama know that?

Do you remember why mileage per gallon was increased in 1975? It was to reduce the price of gasoline; the price had reached four dollars a gallon. During the '70s there had been long lines at gas stations, and prices had increased, The government's increase in mileage for a gallon of gas was designed to force people to drive less, use less gasoline -- and the reduction in demand for gasoline would bring the cost of gasoline down and reduce imports.

Government predictions in 1975 were the same as Obama's – and we know that did not work out well.

The U.S. government's 40-year-old corporate average fuel economy (CAFE) program is a case study in unintended consequences. During its first twenty-five years, CAFE boosted domestic sales of Japanese and European imports, which, [because of their foreign market small-car requirements,] typically had a 50% higher mpg rating than American automobiles in 1975. Partly as a consequence of CAFE, the U.S. market share of foreign-designed vehicles increased from 18% in 1975 to 29% in 1980 and 41% in 2000, according to the National Research Council. In addition, the idea that increased CAFE standards will lead to less oil importing and gas consumption is demonstrably false. "In 1975, before CAFE, we imported 37 per cent of our petroleum requirements," wrote Michael Heberling in 2006. "According to the government's Monthly Energy Review of July, 2005, with CAFE we now import 64 per cent. CAFE neither reduced America's use of foreign oil nor lowered our consumption of gasoline."

Few members of Congress anticipated or desired such disastrous results when they created the CAFE program in 1975.

You would think the history would worry Obama.

In October, 2011, Bob Lutz, former vice chairman of General Motors, talked to the Senior Men's Club of Birmingham. His topic involved past problems between car designers and bean counters during the old days. Problems of the recent past, according to him, were caused by bean counters.

In the discussion period after his talk, I presented my problem. It appeared that automobile manufacturers were designing to satisfy the federal government, instead of appealing to car buyers. All the cars looked alike; they were small, ugly, foreign looking – with the front end down, tail up. I told him that I had just had a new engine installed in my old 1993 Fleetwood, instead of buying a new car. He asked me if I had a question. (I had been wondering for some time why the auto industry didn't object to government requirements.) He then explained: GM could redesign to meet the 54.5 mpg requirement and could compete with foreign builders. (GM had now designed the little foreign automobile-like things.) And he casually mentioned that it would just add about $6000 cost to each automobile.

He can't possibly know the cost. But anyhow I then knew, for the first time, that U.S. automobile manufacturers will acquiesce to anything the government is ignorant enough to propose. Their dealers, however, are concerned; they know the increased prices for automobiles will be disrupting for sales. But the manufacturers? They don't worry. Old habits linger; they have now returned to the old days, with a change or two.

Back then the U.S. automakers obeyed the unions' commands and added to the prices their customers had to pay to satisfy the unions' demands. Now, in a situation not quite as comfortable, they are dancing to the federal government's tune – and still don't realize: To survive in the long run, they should be designing automobiles to satisfy their customers – not some third party, even if it is the federal government.

We know the upped mpg has been estimated to add $6000 in cost to an automobile, but we don't know the resulting price, and we don't know what the price of gasoline will be during the coming years, so how does Obama know there will be a savings of $8000 per automobile?

He doesn't; he is lying, just as he has done so often in the past.

When 2008 and the recession hit, GM and Chrysler squealed for help. Republicans and Democrats were happy to help; after all, the companies were too big to fail. Failure was prevented with the arrival of the government (bearing the people's money), but one big problem persists -- much of the government's regulations and the unions' labor and work-rule costs that caused the problems originally, still remain. Only the passage of time will tell how well the safety net will work. We know that for the first time that I am aware of, preferred stock holders, who were best prepared for the possibility of bad times, did not receive anything for their efforts. The administration decided who would receive the leftover fruits from the saved GM and Chrysler corporations, and it selected labor unions.

Thepoliticalguide.com tells us:

The bankruptcy restructuring plan agreed upon by the government and GM gave the U.S. government a 60% share in the company and gave the Canadian government a 12% share. The United Auto Workers gave up a health and savings plan worth $20 Billion in exchange for a 17.5% share in the company and over $8 Billion in debt and preferred stock. Bondholders held $27 Billion in stock prior to the collapse and received only a 10% equity share in the new GM company.

- Days before GM was placed into bankruptcy, the Obama administration demanded and received the resignation of company CEO, Rick Wagoner.
- The Obama administration pushed for the closing of numerous GM dealerships

- The substance of the bankruptcy settlement was heavily tilted to favor unions - a result that many people suggest would not have occurred without political motivations
- The bankruptcy settlement allowed the U.S. government, the Canadian, and the UAW Union to appoint chairs of the board - an action that would directly change the trajectory of the company for years.

I don't think the future bodes well for GM. And I have always had this problem: Can government ever solve a problem which it created through ignorance, by applying the same brains for remedial work?

Monopolies

The government has always done a good service, we have been told; it has protected us from the ravages of monopolies that threaten us from out-of-control corporations.

We have approximately twenty-nine million firms in the United States, all competing for their share and more of the market. In that environment, monopolies are impossible to form. The government, knowing little about business, has developed laws to protect us from dangerous monopolies. And it has exercised the laws in the past. But if you review the history of all monopoly activities initiated against firms, you will find that in all cases the government acted at the request of the firm's competition. Those charged, in almost all cases, had prices too low, were very successful, and their competition needed help from the government in order to compete. Dangerous monopolies in the private sector have never existed. There are no laws, however, to protect us from government monopolies -- and they are real. (Think Freddie Mac, Fannie Mae – the business they have taken from the private sector, and the nationwide problems that they have fomented.)

CHAPTER **II**

The Financial Marketplace...
Problems that Face Us

AS MENTIONED PREVIOUSLY, we have approximately twenty-nine million firms in the United States, all competing for their share and more of the market, and their individual activities create our marketplace. Their fortunes, collectively, determine whether our nation is prospering or otherwise. The complexity of this marketplace, where services and goods are made and sold, has been introduced prior to this point. Now, let's look at the financial area. There are far fewer financial institutions. In 2011 there were only 7313 FDIC insured institutions – 7303 commercial banks and savings institutions and 10 U.S. branches of foreign banks. What about complexity in the financial sector?

Only 7313 financial institutions? That seems far too few to me; a larger number of smaller banks would be far safer.

That is my opinion, developed from my direct experience with stock investments over the years and new information gleaned lately concerning banking. The Wall Street Journal and John Steele Gordon, October 10, 2008, introduced some knowledge and history into the picture.

We are now in the midst of a major financial panic. This is not a unique occurrence in American history. Indeed, we've had one roughly every twenty years: in 1819, 1836, 1857, 1873, 1893, 1907, 1929, 1987 and now 2008. Many of these marked the beginning of an extended period of economic depression.

How could the richest and most productive economy the world has ever known have a financial system so prone to periodic and catastrophic break down? One answer is the baleful influence of Thomas Jefferson.

Jefferson, to be sure, was a genius and fully deserves his place on Mt. Rushmore. But he was also a quintessential intellectual who was often insulated from the real world. He hated commerce, he hated speculators, he hated the grubby business of getting and spending (except his own spending, of course, which eventually bankrupted him). Most of all, he hated banks, the symbol for him of concentrated economic power. Because he was the founder of an enduring political movement, his influence has been strongly felt to the present day.

Consider central banking. A central bank's most important jobs are to guard the money supply -- regulating the economy thereby -- and to act as a lender of last resort to regular banks in times of financial distress. Central banks are, by their nature, very large and powerful institutions. They need to be to be effective.

Jefferson's chief political rival, Alexander Hamilton, had grown up almost literally in a counting house, in the West Indian island of St. Croix, managing the place by the time he was in his middle teens. He had a profound and practical understanding of markets and how they work, an understanding that Jefferson, born a landed aristocrat who lived off the labor of slaves, utterly lacked.

Hamilton wanted to establish a central bank modeled on the Bank of England. The government would own 20% of the stock, have two seats on the board, and the right to inspect the books at any time. But, like the Bank of England then, it would otherwise be owned by its stockholders.

To Jefferson, who may not have understood the concept of central banking, Hamilton's idea was what today might be called "a giveaway to the rich." He fought it tooth and nail, but Hamilton won the battle and the Bank of the United States was established in 1792. It was a big success and its stockholders did very well.

The First Bank of the United States, 1791-1811, was formed to help the government to work out of the debt left from the Revolutionary War. The Federal Reserve Bank of Minneapolis tells us that the first bank was established with $10 million of capital obtained by selling stock; private investors owned 80% and the federal government owned 20%. The bank helped the government fund the debt, and it issued bank notes or currency. The first bank, in addition to the one in Philadelphia, had eight other branches, located in the major cities of those days.

It provided the country with a regular money supply with its own banknotes, and a coherent, disciplined banking system.

But as the Federalists lost power and the Jeffersonian became the dominant party, the bank's charter was not renewed in 1811. The near-disaster of the War of 1812 caused President James Madison to realize the virtues of a central bank and a second bank was established in 1816. But President Andrew Jackson, a Jeffersonian to his core, [did not renew it in 1832, it died in 1836,] and the country had no central bank for the next seventy-three years.

We paid a heavy price for the Jeffersonian aversion to central banking. Without a central bank there was no way to inject liquidity into the banking system to stem a panic. As a result, the panics of the 19th century were far worse here than in Europe and precipitated longer and deeper depressions. In 1907, J.P. Morgan, probably the most powerful private banker who ever lived, acted as the central bank to end the panic that year.

Even Jefferson's political heirs realized after 1907 that what was now the largest economy in the world could not do without a central bank. The Federal Reserve was created in 1913. But, again, they fought to make it weaker rather than stronger. Instead of one central bank, they created twelve separate banks located across the country and only weakly coordinated.

The Federal Bank of Minneapolis adds: "There was concern that the new central bank would be run by and for Wall Street, and so it was important to the founders that the bank not be focused on New York. Thus the system was decentralized into [12] District Banks, which operated independently, and with an oversight board located in Washington, D.C. Each district bank issued its own money, backed by the promise to redeem this money in gold."

No small part of the reason that an ordinary recession that began in the spring of 1929 turned into the calamity of the Great Depression was the inability of the Federal Reserve to do its job. It was completely reorganized in 1934 and the U.S. finally had a central bank with the powers it needed to function. That is a principal reason there was no panic for nearly sixty years after 1929 and the crash of 1987 had no lasting effect on the American economy.

While the Constitution gives the federal government control of the money supply, it is silent on the control of banks, which create money. In the early days they created money both through making loans and by issuing banknotes and today do so by extending credit. Had Hamilton's Bank of the United States been allowed to survive, it might well have evolved the uniform regulatory regime that a banking system needs to flourish.

Without it, banking regulation was left to the states. Some states provided firm regulation, others hardly any. Many states, influenced by Jeffersonian notions of the evils of powerful banks, made sure they remained small by forbidding branching. In banking, small means weak. There were about a thousand banks in the country by 1840, but that does not convey the whole story. Half the banks that opened between 1810 and 1820 had failed by 1825, as did half those founded in the 1830s by 1845.

Many "wildcat banks," so called because they were headquartered "out among the wildcats," were simple frauds, issuing as many banknotes as they could before disappearing. By the 1840s there were thousands of issues of banknotes in circulation and publishers did a brisk business in "banknote detectors" to help catch frauds.

The Civil War ended this monetary chaos when Congress passed the National Bank Act, offering federal charters to banks that had enough capital and would submit to strict regulation. Banknotes issued by national banks had to be uniform in design and backed by substantial reserves invested in federal bonds. Meanwhile Congress got the state banks out of the banknote business by putting a 10% tax on their issuance. But National banks could not branch if their state did not allow it and could not branch across state lines.

Unfortunately state banks did not disappear, but proliferated as never before. By 1920, there were almost 30,000 banks in the U.S., more than the rest of the world put together. Overwhelmingly they were small, "unitary" banks with capital under $1 million. As each of these unitary banks was tied to a local economy, if that economy went south, the bank often failed. As depression began to spread through American agriculture in the 1920s, bank failures averaged over 550 a year. With the Great Depression, a tsunami of bank failures threatened the collapse of the system.

The reorganization of the Federal Reserve and the creation of the Federal Deposit Insurance Corporation hugely reduced the number of bank failures and mostly ended bank runs. But there remained thousands of banks, along with thousands of savings and loan associations, mutual savings banks, and trust companies. While these were all banks, taking deposits and making loans, they were regulated, often at cross purposes, by different authorities. The Comptroller of the Currency, the Federal Reserve, the FDIC, the Federal Savings and Loan Insurance Commission (FSLIC), the Securities and Exchange Commission (SEC), the banking regulators of the states, and numerous other agencies all had jurisdiction over aspects of the American banking system.

The system was stable in the prosperous postwar years, but when inflation took off in the late 1960s, it began to break down. S&Ls, small and local but with disproportionate political influence, should have been forced to merge or liquidate when they could not compete in the new financial environment. Instead Congress made a series of quick fixes that made disaster inevitable.

In the 1990s interstate banking was finally allowed, creating nationwide banks of unprecedented size. But Congress's attempt to force banks to make home loans to people who had limited creditworthiness, while encouraging Fannie Mae and Freddie Mac to take these dubious loans off their hands so that the banks could make still more of them, created another crisis in the banking system that is now playing out.

While it will be painful, the present crisis will at least provide another opportunity to give this country, finally, a unified banking system of large, diversified, well-capitalized banking institutions that are under the control of a unified and coherent regulatory system free of undue political influence.

Mr. Gordon writes with clarity; he is the author of *An Empire of Wealth: The Epic History of American Economic Power* (HarperCollins, 2004).

Theories of central banks and how they should work abound. But it should be obvious – absolute knowledge in this financial area does not exist.

Our economic problems, still in place in 2012, first became apparent in the financial sector in December, 2007. Home mortgages and too many failures of buyers to meet their obligations began to make the news, and that signaled the beginning of the downturn. But before we assign blame to home buyers and our old problems with central banking, let's look at the whole picture; something else had to be on the scene: There is no way our house-buyer failures could have caused a downturn in the European Union and other parts of the world.

The look was difficult, because the financial people have a jargon all their own, and I had to wade a little in order to understand. But I did it, because I think we need to know who to blame – who the culprits were, and what they did to cause all this grief. I looked for clarity; when I didn't find it, I looked for clarifiying definitions of the words or phrases.

(Something I have learned: When information for public consumption is presented with words most of the public does not understand, it generally means that those who wrote them were quoting someone else and didn't know the subject either.)

A Beginning Financial Problem

Jeffrey Friedman provided much information on the cause of our 2007 economic downturn and why it went worldwide in a January/February 2010 Cato Institute Policy Report. His work, in words that I could understand, was instrumental in clarifiying many details.

The financial problem, still an affliction, was not caused by corporations, banks or financial institutions; they were caused by government – its regulations and mistakes -- and we will cover them in detail. These were important events:

1) Federal Deposit Insurance Corporation (FDIC), established in 1933, to insure bank deposits
2) Banks forced to hold financial cushion, 10% of outstanding loans; (FDIC protection could make them reckless)
3) Home mortgages begin to be used by security firms as assets, for mortgage-backed securities (stocks)
4) The mortgages of borrowers with poor credit ratings begin to be used for securities, pay higher dividends
5) Housing bubble – houses begin to increase in value
6) Recourse rule: Smaller financial cushion for AAA, AA rated mortage-based securities (MBS)
7) Three rating companies, which rated the stocks and bonds, made mistakes
8) International acceptance of AAA, AA rated MBSs
9) Subprime mortgages begin to fail, borrowers couldn't meet obligations

The foundation for the 2007 problem was established in 1933. That year the U.S. government formed the Federal Deposit Insurance Corporation (FDIC). It was developed to insure people's bank savings; if an FDIC associated bank failed, its depositors were assured that their money was safe. If necessary, any loss would be replenished from federal government coffers. FDIC was established to prevent bank closings and panics during a business downturn. People had lost money when the economy had faltered and banks had closed during the 19th century. In the 20th century, in 1933, an estimated 4000 banks had failed in the United States.

But after making the depositors' money safe, the FDIC worried that the banks would use the safety of that insurance to make riskier loans than they would have made otherwise. The FDIC, with its insurance against losses, had removed Fear of Failure from banking operations, a recognized regulator of all businesses – even at that early date. So, the FDIC substituted government controls: Laws to force banks to hold a cushion (for example, money) against potential losses. The banks had to maintain a government-specified percentage of the total amount of money it had loaned, set aside and readily available if needed, for any problem that became evident. Overseas governments instituted safety measures for their banks, also.

Vern McKinley tells us In *Financing Failure: Books: The Independent Institute*, about the first too big to fail (TBTF) incident:

> ...In the early 1950s, Congress phased out the [Reconstruction Finance Corporation (RFC) developed by Hoover in the '30s, serving the same purpose] and gave the Federal Deposit Insurance Corporation (FDIC) new powers to assist troubled banks, including the authority to bail out creditors and shareholders of banks in danger of closing, a power not used until 1971. The FDIC faced one of its greatest challenges in 1974 when Franklin National Bank of New York, one of the twenty largest U.S. banks, experienced a massive run and lost nearly 50 percent of its deposits. The FDIC, the Federal Reserve, and the Office of the Comptroller of the Currency reacted by arranging a bailout of uninsured creditors and depositors. Federal regulators worried that without a bailout, financial markets would have been severely disrupted, but no regulator

has ever released a detailed analysis of the expected disruption. The Franklin resolution marked a turning point: it was the first bailout of an institution deemed "Too Big to Fail," and it foreshadowed the method of future bank resolutions....

Mr. Friedman, continuing, explained international efforts.

...In 1988, financial regulators for G10 countries agreed to Basel I Accords, an effort to standardize the world's bank-capital regulations to assure safety for depositors' funds, and it was subsequently accepted by most of the civilized world. The banks did not have to devote any capital to its holdings of government bonds, cash, or gold — the safest assets. But they had to allot four percent of their capital [as a financial cushion] to each mortgage that it issued and eight percent to commercial loans and corporate *bonds* [in case too many borrowers failed to pay as promised] ....

(In explanation of *bonds*, just mentioned, all documented <u>contracts</u> and <u>loan agreements</u> are bonds. When you, a business, or a corporation borrow from a bank, and you engage in a written and signed <u>promise to pay</u> a certain <u>sum</u> of <u>money</u> on a certain date, or on <u>fulfillment</u> of a specified <u>condition</u>, you create a bond. And that applies to businesses of all sizes.)

...Countries implemented the regulations, with their own variables, and in 1991, the United States joined in and, essentially, ruled that a 10% cushion was required for "well capitalized" commercial banks....

Mortgage-Backed Securities Arrive

A principle service of banks, early on, of course, was to make money available to people who wished to buy homes. The people agreed to the bank's terms and agreed to pay a specified percentage of the total amount borrowed plus interest each month. The bank held the loan, and the buyer dealt with the bank until the debt was paid.

The 10%-of-all-loans-backup requirement was a financial problem for the banks that then existed. Some bank funds were tied up during the entire life of the loan – and that could be thirty years. And with many loans outstanding, a substantial amount of capital could be involved. Busy minds went to work. A potential bank solution arrived – why not take a bunch of its loans, arrange them into a package, and offer the collection for sale as a security?

That was done, and the mortgage-backed security (MBS) was born. A bank makes a loan to an individual for a house, as described above, and either holds it, combines it with other loans with the same interest and maturities to form a security, or sells it to another company for the latter purpose. The security is then sold to investors who receive the interest and principal payments as dividends as monthly loan payments are made. As the house buyer makes payments to the bank, the bank keeps a fee or charge and sends the rest to the MBS owner. The MBS owner keeps a fee and passes the rest of the principal and interest payments to the investors.

The mortgage-backed-security market brings more funding into the mortgage industry, making it possible to finance more loans for residential or commercial real estate. In the beginning, MBSs involved mortgages to prime borrowers (those with good credit records) and mortgages with governmental guarantees. But beginning in about 1997, investors became intrigued by securities involving loans to borrowers with poor credit histories. These high-risk borrowers were forced to pay higher interest rates, for example, and the securities involving the higher-risk loans (subprime MBSs) paid greater dividends than did those involving safer borrowers. That factor established the market for subprime MBSs.

Subprime loans and associated securities grew immensely because of government pressures. The government drive was on to increase home ownership, the pressure on banks to lower their standards for credit was heavy. People who were incapable of handling their mortgages were buying houses because of relaxed standards, and some were buying more than one house -- to resell at a profit. The market for houses was expanding greatly.

As time passed and securitization gained momentum, the activity of banks that originated loans changed; some made the loan, sold it, and no longer had any interest. Certain lenders specialized in sub-prime mortgages, but most of them only originated the mortgages, passed them on to security companies, the security companies formed the securities, and the originating banks had no further interaction with the loan or the securities formed; the loans were completely eliminated from their portfolios.

Thus, the loan originators, after the sale to the security company, had no further interaction with the loan or investor, and the security company was in complete charge. All dangers from risky loans were now eliminated from the loan-originating bank. Some accusing fingers, appropriately, have pointed at the loan originators: they knew more about the mortgages that failed, and the risk that caused them, than did those who formed the securities.

Another Step in the Downturn Cause

The regulation, an amendment to Basel I that furthered the grief was issued in 2001. The FDIC, the Fed, the Comptroller of the Currency, and the Office of Thrift Supervision instituted the Recourse Rule, which changed risks: Asset-backed securities (ABS) such as bonds backed by credit card debt, car loans, or mortgages required a mere two percent capital cushion, as long as they were rated AA or AAA or were issued by a government-sponsored enterprise (GSE), such as Fannie Mae or Freddie Mac. Thus, where a well-capitalized commercial bank needed to devote $10 of capital to $100 worth of commercial loans or corporate bonds, or $5 to $100 worth of mortgages, it needed to hold only $2 of capital on a mortgage-backed security (MBS) worth $100.

...In 2006, the Bank for International Settlements followed suit with their own, similar backing of AA and AAA mortgage backed securities. In 2006, Basel II began to be implemented outside the United States, and it took the Recourse Rule's approach -- encouraging foreign banks to stock up on GSE-issued or highly rated MBSs. Thus, overseas participants entered the fold....

...According to an array of scholars from around the world — Viral Acharya, Juliusz Jablecki, Wladimir Kraus, Mateusz Machaj, and Matthew Richardson — these regulations helped turn an American housing crisis into the world's worst recession in seventy years...."

And Then...

The FDIC, the Fed, the Comptroller of the Currency, and the Office of Thrift Supervision put their trust in three rating companies when they placed the responsibility for determining bond ratings on their shoulders. They did this because they thought the rating companies were subject to "market discipline." That was a mistake; the rating companies were created by the security exchange commission (SEC), and in 1975 it had established the three rating companies, S&P, Moody's and Fitch, as legally protected entities – with no fear from any competition that had, previously, made their work reliable. And, as Mr. Friedman further points out, "... these three 'rating agencies' had gotten sloppy. Moody's did not update its model of the residential mortgage market after 2002, when the boom was barely underway...."

...By 2008 approximately 81 percent of all the rated mortgage-backed securities held by American commercial banks were rated AAA, and 93 percent of all the MBSs that the banks held were either triple-A rated or were issued by a GSE, thus complying with the Recourse Rule. (Figures for the proportion of double-A bonds [were] not yet available at publication time.) According to the scholars I mentioned earlier, the lesson is clear: the commercial banks loaded up on mortgage backed securities because of the extremely favorable treatment that they received under the Recourse Rule -- as long as they were issued by a government sponsored enterprise (Freddy Mac or Fanny Mae), or were deemed safe by the government's rating agencies, Fitch, Moody's, or S&P, with AA or AAA ratings.

When subprime mortgages began to default in the summer of 2007, however, those high ratings were cast into doubt. A year later, the doubts turned into panic. Federally mandated mark-to-market accounting — the requirement that assets be valued at the price for which they could be sold right now — translated temporary market sentiment into actual numbers on a bank's balance sheet, so when the market for MBSs dried up, Lehman Brothers went bankrupt — on paper. Mark-to-market accounting applied to commercial banks too. And it was the commercial banks' worry about their own and their counterparties' solvency, due to their MBS holdings, that caused the lending freeze and, thus, the Great Recession.

Mark-to-market accounting. If you have done any investing in stocks, you have heard: If stock prices go down, don't panic -- they can go up again -- you don't lose anything until you sell. Well, that didn't apply to Lehman and other financial institutions. Mark to market meant that when their assets decreased in value, their financial statement reflected that directly, and if panicked clients caused prices to go down very fast and low -- even though banks didn't intend to sell – and liabilities exceeded assets they were ruled through by regulations -- broke. Lehman Brothers, during bankruptcy procedures, blamed much of its third quarter 2008 loss on mark-to-market accounting rules.

The Rating Agencies

Halah Touryalai, Forbes Magazine, in her article, *S&P, Moody's, and Fitch: Do We Need Them?* raises a question or two about rating agencies. How dependable, for the future, are agencies that rated mortgage-backed securities as AAA or AA, and safe – which obviously were not.

Imagine if Lehman Brothers survived the financial crisis in 2008, and in 2011 was advising clients to buy large positions in subprime mortgages. Better yet, imagine that the requirements for subprime mortgages remained the same: no income, no job and no assets – still, you got a loan.

Sounds crazy, doesn't it? So then why do the agencies that blessed toxic assets with AAA ratings for several years leading up to the crisis still carry so much influence in the markets?

Forget the crisis. Think about the way these agencies are paid: The companies and governments they rate pay them. That's like a student paying his teacher for a report card. It should be enough to send investors running in the opposite direction when they see a new credit rating.

Moreover these ratings agencies are the same ones that were dubbed "key enablers of the financial meltdown" by the U.S. Financial Crisis Inquiry Commission 2011. They're also the same agencies that made a $2 trillion error when downgrading the credit rating of the United States.

http://www.forbes.com/sites/halahtouryalai/2011/12/06/sp-moodys-and-fitch-do-we-need-them/

Government can only react to a financial problem, and past history tells us it then solves it incorrectly. And the failures apply to all banks – nationwide -- causing far more problems for the economy.

When government took away the fear of failure from banks and companies, government lost its way.

Government, or any other designated group or individual, cannot regulate practices that will be good for the majority of 7313 financial and 29,000,000 other types of firms. What is good for the goose's business frequently harms the gander's.

States Helped

The states helped make the problem worse with the prevalence of "no-recourse" laws: People who couldn't afford the houses they bought could merely walk away from them -- without any financial liability. And the elimination of risk entered the picture again. Many people took on unaffordable mortgages with virtually no risk. Mr. Friedman criticized the 1933 error that helped create this economic mess.

> ...The theory behind deposit insurance was (and remains, still) that banking is inherently prone to bank runs, which had been common in 19th-century America and had swept the country at the start of the Depression.
>
> But that theory is wrong, according to such economic historians as Kevin Dowd, George Selgin, and Kurt Schuler, who argue that bank panics were almost uniquely American events (there were none in Canada during the Depression — and Canada didn't have deposit insurance until 1967). According to these scholars, bank runs were caused by 19th-century regulations that impeded branch banking [did not allow banks to distribute their activities to smaller neighborhood institutions] and bank "clearinghouses" [where private banks agreed to pool resources, form a clearing house, and would perform as lenders during banking calamitous periods; in addition, the association (clearing house) audited banks periodically to assure their stability]. Thus, deposit insurance, hence capital minima, hence the Basel rules might all have been a mistake, founded on the New Deal legislators' and regulators' ignorance of the fact that panics, like the ones that had just gripped America, were the unintended effects of previous [government] regulations....

In the first quarter of 2009, Congress changed the mark-to-market accounting law – the tremendously disastrous one that had forced Lehman Brothers to close. The Financial Accounting Standards Board (FASB) approved new guidelines; the valuation of bank assets in the future is to be based on a price that would be received in an orderly market, rather than one determined by a present market's disorderly situation.

This treatise only touches on the problems; read the entire article: *A Perfect Storm of Ignorance* http://www/cato/org/pubs/policy_report/ v32n1-1.html.

Let's Minimize Economic Downturns

For years I paid little attention to financial activities; I didn't understand them. I would start an article, run into several words or terms foreign to me, and I would quit. I wasn't too interested, anyhow. A year or so ago, in the midst of the current downturn, I decided to read them for understanding -- and when I ran into financial jargon or buzzwords foreign to me, instead of quitting, I began to check them out with an extremely handy Internet tool, Google. With that added attention, I soon realized why I didn't understand. The complexity of the manufacturing and services marketplace, mentioned earlier, is far worse in the financial arena. There is little understanding possible or available. Theories abound, but absolute knowledge is missing – and not because of ignorance. Unfortunately, there is no absolute knowledge possible – the problems are too complex.

For solutions, the Founders eliminated banking from the federal government's grasp; they intended the state governments to face the problems and to enlist the help of the people. But government, instead, changed the rules to allow them to face the problems and to solve them. Since our beginnings, financial problems perceived by government have been resolved by government with regulations. Laws passed centuries ago remain on the books. The laws didn't solve any problems then, and they won't solve any in the future -- but they remain on the books. I now know that theories abound about financial activities and problems, but knowledge and solutions to them are missing.

And this is what Mr. Friedman says about that:

> ...Omniscience cannot be expected of human beings. One really would have had to be a god to master the millions of pages in the Federal Register — not to mention the pages of the Register's state, local, and now international counterparts. No one could pick out the specific group of regulations, issued in different fields over the course of decades, which would end up conspiring to create the greatest banking crisis since the Great Depression. This storm may have been perfect, therefore, but it may not prove to be rare. New regulations are bound to interact unexpectedly with old ones if the regulators, being human, are ignorant of the old ones and of their effects.

> Get ready for some new ones. The SEC's response to the crisis has not been to repeal its 1975 regulation [which removed Fitch, S&P, and Moody's from competitive raters], but to promise closer regulation of the rating agencies. And instead of repealing Basel I or Basel II, the Bank for International Settlements (BIS) is busily working on Basel III, which will even more finely tune capital requirements and, of course, increase capital cushions....

The financial cushion, forced on the banks to prevent them from doing risky things and to assure depositors of the safety of their money, as Mr. Friedman, et al., have shown, did not help; it did nothing to prevent the problems that occurred:

> ...The aggregate capital cushion of all American banks at the start of 2008 stood at 13 percent — one-third higher than the American minimum, [and] one-fifth higher than the Basel minimum. Contrary to the regulators' assumption that bankers

need regulators to protect them from their own recklessness, the financial crisis was not caused by too much bank leverage but by the form it took: mortgage-backed securities. And that was the direct result of the fine tuning done by the Recourse Rule and Basel II....

The administration and its cronies have the microphone and the popular media's attention. Government was the culprit, but the popular media have not reported that, and we have had to do a little work with dependable media sources and the Internet to discover who the real culprits were.

But There Were Other Problems

Some things were going on that few of us in the public arena realized, and they demonstrate the differences between financial institutions and the other side of business. It is very difficult for a manufacturing operation to hide its economic condition; its quarterly report provides sales income, costs, taxes, resulting profit -- and the details are difficult to hide. Some games are played, such as the distribution of sales, for instance. Some CEO might assign certain of this quarter's sales and income to next quarter's for better appearances or vice versa. But the amount of possible finagling is quite small. Not so, you will find, in banking and finance circles.

Banking and financial activities are not as visible. The product is money. Conventional banking with deposits, loans for housing, automobiles, and other things are quite understandable. But when we get involved with investments and the stock market, problems arise.

There is no visible product – unless you consider a pile of banknotes or certificates a visible product. And the reports of some of those financial activities are described by the media in words that only individuals heavily involved with those financial affairs could understand. How about capital arbitrage? Or structured investment vehicles (SIV)? Or off-balance-sheet ventures, shadow institutions, the repo market, or Credit default swaps (CDS)? Let's look at some of these now, and we will get an inkling of the complexity involved in this area of endeavor.

But first you should know that "Basel I included provisions for very profitable forms of 'capital arbitrage' through 'off-balance-sheet entities' such as 'structured investment vehicles' -- heavily used in Europe and in the United States beginning in about 2000." Wow!

The dictionary definition of arbitrage is sufficient to alert anybody to some potential complexities: "Arbitrage: the simultaneous purchase and sale of the same securities, commodities, or foreign exchange in different markets to profit from unequal prices." How about that?

Let's try again. The institutions that engage in capital arbitrage use short term borrowing at low interest rates, to engage in long term lending at larger interest rates or fees. For example, I can borrow funds for a short period of time at a low rate of interest, and if times are good and I need more, I can get more. So I gamble on good times continuing; I borrow short-term money from a pension fund at a low interest rate and use it to buy some mortgage based securities whose rate of return is greater than the interest I am paying. And I prospered until the bad times began. People began to fail to pay their mortgages, and my returns began to suffer. Then, too many people failed, times got worse, the security became worthless. I owed the bank I was borrowing from, I could borrow no money from any other source (had insufficient capital) with which to pay, so I crashed, I couldn't pay the pension fund, and that brought hard times to the fund.

These activities are not conventional banking activities, so the normal regulations do not apply; no financial cushions are required if problems arise – they are off-balance-sheet entities. They are risky activities, with little or no capital involved for the perpetrators, so during good times they can generate significant profits -- even in relatively low-margin businesses. However, during times of stress, such as those in 2007, the power of leverage works in reverse. The negative impact is amplified, and substantial capital (money) is required, but may not be available. These are structured-investment vehicles.

There was a way that I could have delayed my crash with the pension fund. I could have joined the repo market. As the time arrived for quarterly reports, I could have reached an agreement with a cash-rich corporation to sell them some different bonds or stock that I owned, worth $105 per share, for a short time, maybe a day or two, which I would buy back at a smaller price. I would sell them the stock for $100, pay the proceeds to the pension fund, reducing my debt in time for the balance-sheet report, then borrow money to buy the stock back for $105. That increased my debt by a little more, but my balance sheet won't show it for about three months, at which time it must be made public. Actually, I was borrowing for a short time to help my report look better, and accounting rules made me look better for report day – a most important day.

The Repo Market

Jacob Goldstein's information was the basis for the previous example. http://www.npr.org/blogs/money/2010/03/repo_105_lehmans_accounting_gi.html

He says:

As the financial crisis grew in 2007 and 2008, Lehman knew it needed to reduce its reliance on borrowed money. But it was a bad time to sell stuff off and pay back debts. So Lehman made special use of something called the repo market.

The deals are short term — the bank often buys back the asset just days after it sells it. During the boom, there was about $12 trillion (with a t!) loaned out in this market at any given time, Yale's Gary Gorton told me...

...When Lehman Brothers wanted to make it look like it wasn't borrowing so much money, the company used [the] special technique.... It did repo deals where it took slightly less cash than the asset was worth.

For example: If Lehman owned a bond that was worth $105, it would "sell" it on the repo market for $100. (The "105" in Repo 105 refers to the fact that the assets were worth at least 105% of what Lehman was getting for them.)

This gap allowed the company to record the transaction as if it had been a true sale of the bond — despite the fact that, under the agreement, the company would repurchase the bond just a week or so after it had sold it.

Lehman would take the money it got from selling the bond and pay off some of its debts. Then, after it had issued its quarterly report, the company would borrow more money to repurchase the bond.

Lehman went big on this technique: In the second quarter of 2008 it used Repo 105 to move $50 billion off of its balance sheet, according to the Examiner's report. "Lehman did not disclose its use ... of Repo 105 to the government, to the rating agencies, to its investors, or to its own Board," the report said. One senior official inside the company warned that the use of Repo 105 would present "reputational risk" to the company if the public found out.

Conventional, commercial banks also ventured into structured investment vehicles, such as those described from time to time, using their money market accounts or other assets.

Do you know the difference between an investment bank (like Lehman) and your commercial bank where you do your checking and saving?

U.S. news answers this question and many more such as this at: http://money.usnews.com/money/blogs/the-home-front/2008/09/19/lehman-brothers-and-your-bank-deposits.

Investment banks operate differently from commercial banks

and thrift institutions. Their primary purpose is to facilitate the sale of stocks and bonds. These Wall Street firms operate as advisers and agents for companies that want to raise capital, often by issuing more stock or other securities.

Commercial banks and thrift institutions take deposits for checking and savings accounts from consumers and businesses. These deposits are insured by the FDIC for up to $100,000 per depositor per insured bank and up to $250,000 for retirement accounts. These banks lend this money to consumers and companies for autos, homes, business equipment, etc.

Credit Default Swaps

I could have bought some protection against loss of my mortgage-backed securities. Thepressnet.com explains.

Allow me to teach you what a credit default swap is and why it's so important to what is happening to the economy today.

Virgle Kent borrows $50 from me. I want to get insurance on his debt in case he goes broke. I go to Roissy and say, "Hey, Virgle Kent owes me $50. Can you insure that debt?"

"I'll insure it if you pay me $4 a year," Roissy says.

"Done!"

Roissy is betting that VK will pay me back, especially since he did his homework by looking at VK's credit rating and saw it was superb. Roissy wrote me a credit default swap, an unregulated derivative <u>invented in 1995 by JP Morgan</u>.

Unfortunately Roissy has some problems with his business, and he no longer even has $50 to pay me in case VK goes broke. The premiums I gave him are long gone. Credit agencies notice this and tell Roissy to find some cash or his credit rating goes down. Roissy is troubled because if his credit rating goes down then he won't be able to raise cash at good rates to keep his business open (today's large businesses need a constant flow of credit to maintain operations). Sure enough his rating gets killed and Roissy goes bankrupt.

Now I'm in trouble. The debt I had on my books that was insured is now uninsured. The agencies look at my books and see I have this exposed debt and they downgrade [me]. I have no choice but to enter bankruptcy as well. But I happened to be knee deep in the Credit Default Swap game too. I wrote a ton of them for Arjewtino, insuring the debt owed to him by other parties. When I go down it puts pressure on him. Like dominoes we fall.

In the carnage it turned out that the ratings we used to judge each other's debt worthiness was bogus from the start. [For a little while even Lehman's record looked good.] Essentially we all gambled like we would at a blackjack table, but we did it while drunk. And blind.

The insurance company AIG wrote $78 billion worth of swaps.

Ivy League Masters of Business Administration (MBAs) turned the default swaps into an even more insidious device. In ways that I will not begin to understand, swaps were used not just to insure against debt but to speculate if companies would fail or not. It turned out that while Virgle Kent only owed me $50, there were swaps written worth $500 between parties that VK didn't even know about! The swaps became a means to make money instead of a simple insurance policy. This was enabled by a government run by politicians whose treasure chests were stocked full of kind donations from the big bankers. They did not hesitate to look the other way.

A lot of swaps were written by banks and businesses that are now very sick from making bad bets and possibly outright fraud in the housing boom. (Who would have thought that giving no money down / no-doc loans was a bad idea?)

Here's the bad news:

...there are $45 trillion of credit default swaps out there. A default on a mere 10% would cause an economic disaster. Unfortunately, it's guaranteed to happen.

Actually that was the good news. Here's the real bad news:

The Bank for International Settlements recently reported that total derivatives trades exceeded one quadrillion dollars – that's 1,000 trillion dollars. How is that figure even possible? The gross domestic product of all the countries in the world is only about 60 trillion dollars. The answer is that gamblers can bet as much as they want. Read more at:

http://thepressnet.com/2012/02/17/what-is-a-credit-default-swap-and-why-we-should-be-trembling-right-now/

"The shadow banking system makes up 25 to 30 percent of the total financial system, according to the Financial Stability Board (FSB), a regulatory task force for the world's group of top 20 economies (G20). This largely unregulated sector was worth about $60 trillion in 2010, having grown from an estimated $27 trillion in 2002, according to the FSB. While the sector's assets declined during the global financial crisis, they have since returned to their pre-crisis peak."

Information from: http://en.wikipedia.org/wiki/Shadow_banking_system
The difficulty with shadow banking is: It is dependent upon stability of the marketplace. In good times they do well. But when an

unexpected downturn occurs, they crash. In addition, market knowledge about these approaches is limited; the public knows little about SIVs or the other legal entities, and investors were caught off guard by the losses. I spent about twenty hours searching for my own understanding, still quite limited, but presented here.

We should all know: Shadow banking institutions take on risks that mainstream banks are unwilling and not allowed to take. But mainstream banks do participate in shadow banking, which avoids much regulation.

And in good times, returns are good, but the safety of our money is dependent upon high speed computers and people that are betting they know what they are doing and that good times will continue.

The Problem

It should be obvious at this point: Our government is lost, and has been lost for generations, in understanding our economy. And they are not, as our Founders warned, equipped to run it.

The actions of government have caused every economic problem that has arisen and spread nationwide in the last 193 years. From 1819 to 2007 there have been twenty-two downturns, and in every instance, government activities were the cause. Government, in efforts to save the populace from potential actions of greedy businessmen, have been the cause of all the nation's major problems. Government was devising regulations to stop problems that had happened (that they had caused) -- to prevent the same thing from ever happening again. Or, in some cases, regulations to prevent potential problems – problems that might happen if it did not control business peoples' activities. In all cases, it was either attempting to favor one segment of the business sector, or to save some segment of the population from losses due to missteps due to potentially risky ventures by business leaders.

Introducing the Federal Reserve System

Before government entered the central banking activities, bearing citizens' tax money, butnowyouknow.net tells us a banking problem developed in 1907 and "...J.P. Morgan convinced various wealthy New York bankers to act as a private Federal Reserve, shoring up the banking system, and halting this economic failure...." As a result, "The Federal Reserve System (also known as the Federal Reserve, and informally as the Fed), the central banking system of the United States, was created on December 23, 1913...." And the message from butnowyouknow.net continues with the advice that we: "Note that [private interception] worked better than any government-mandated monetary rescue before [1907] or since, including those by the Federal Reserve, FDIC, or the recent $700,000,000,000 bailout."

You may wish to review the complete article; go to: http://butnowyouknow.net/those-who-fail-to-learn-from-history/history-of-economic-downturns-in-the-us.

There Are Better Ways

In 1933 the U.S. government formed the Federal Deposit Insurance Corporation (FDIC). As indicated previously, it was developed to insure people's savings; if an FDIC associated bank failed, its depositors were assured that their money would be safe. Any losses would be replenished from federal government coffers. FDIC was established to prevent bank closings and panics during a business downturn.

Government has no money of its own, and to fulfill its promise, we have seen a lot of private citizens' money go down the drain, used to replenish the coffers of other citizens who had suffered losses. In fact, we are paying now for a very recent problem, so we should all be interested in some reasonable plan which doesn't involve government and our money.

The Solution

The Wall Street Journal reported one in its Opinion Page, *Notable & Quotable*: March 12, 2012. Fear of failure, as indicated previously, tends to keep managers' heads on straight. And one bank agrees; it has used a solution aimed at its management, which has been highly successful and gets the government out of the picture.

Brown Brothers Harriman, Wall Street's oldest surviving general partnership bank, has always used a method for assuring the safety of their clients' money. James Grant, in a speech to the New York Federal Bank, presented a solution developed from the experience of his bank's operations. It was using an approach built on fear.

> [W]hat makes a good banker is more than skill. It is also the fear of God, or, more specifically, accountability for the solvency of the institution that he or she owns or manages.

> To stay out of trouble, the general partners of Brown Brothers Harriman, Wall Street's oldest surviving general partnership, need no regulatory pep talk. Each partner is liable for the debts of the firm to the full extent of his or her net worth. My colleague, Paul Isaac, who is with me today--doubling as my food and beverage taster—has an intriguing suggestion for instilling the credit culture more deeply in our semi-socialized banking institutions.

> We can't turn limited liability [companies, identified by LLC after their names; their owners have limited personal liability for the debts and actions of the LLC.] into general partnerships. Nor could we easily reinstate the so-called *double liability law* on bank stockholders.

What is the *Double Liability Law*?

> Chris Kitze, beforeitsnews.com, provides Yale Law School Legal Scholarship Repository's information on this subject: "...For three quarters of a century - between, roughly, the Civil War and the Great Depression--shareholders in American banks were responsible not only for their investments, but also for a portion of the bank's debts after insolvency. If a bank failed, the court receiver would determine the extent of the insolvency and then assess [officers,

directors, and] shareholders for an amount [necessary to cover losses]....

And he provides an example from those early days:

"...P.J. Kemmeter was one of the leading businessmen in Granton, Wisconsin, owning a heading mill, a general store and a home on one entire city block. He was also a share-holder in the Granton Farmer's State Bank....

"...According to the local newspaper: 'The Farmers State Bank probably does the biggest business of any bank in the state with a capital stock of $10,000. The deposits range from $175,000 to $200,000 and the business done for the wealthy farmers in the vicinity would put to shame many an institution which is more pretentious in the larger cities.

"'...Mr. Kemmeter had invested $200 in the total capital-ization at the bank, which was $10,000, shown by a bank balance sheet in 1912, long before the collapse....

"'...Unfortunately, this bank, swept up in the depression of 1930, was declared insolvent by bank examiners. Shortly af-ter the bank went under, P.J. Kemmeter, a stockholder, [went under] as well....."

"...From the Clark County News, May, 1930: 'M. E. Wilding, trustee in bankruptcy, last week disposed of the P. J. Kemmeter property in Granton. The store building was sold for $3,000 and the home for $2,250, both purchased by John Pietenpol. [A] clock and a tea cart are all that was left from his estate after the bankruptcy and auction on his front lawn."

"...The regime of bank double liability was rejected and abandoned on three grounds: (1) that it had failed to protect bank creditors; (2) that it did not maintain public confidence in the banking system; and (3) that deposit insurance was a far preferable means for accomplishing the regulatory objectives. These arguments have a certain irony in light of recent history, in which deposit insurance itself has proved incapable of protecting the soundness of the banking system or maintaining public confidence in the nation's depository institutions. Double liability, on the other hand, holds the promise for instilling sound banking practices through the application of incentives and shareholder monitoring, rather than the pervasive regulatory scrutiny necessitated under deposit insurance systems. History shows that the nation took a wrong turn when it abandoned double liability for a system of governmentally administered deposit insurance....

"...Bank officers, directors and investors took this responsibility very, very seriously. This article was written in 1994, following the previous worst banking crisis since the Great Depression, the Savings and Loan debacle, and outlines the reasons why it should still be in use today...."

(Read the entire article: http://beforeitsnews.com/banksters/ 2012/02/double-liability-for-bank-shareholders-officers-and-directors-1720343.html)

Paul Isaac resumes his proposal for assuring the safety of bank deposits:

But what we could and should do, Paul urges, is to claw back that portion of the compensation paid out by a failed bank in excess of ten times the average wage in manufacturing for the seven full calendar years before the ruined bank hit the wall. [After a bank fails, the compensation paid to its managers should be calculated for seven full years before failure; that amount should then be compared to ten times the average wage in manufacturing for those same seven years, and the managers should be required to repay any amount that exceeds that sum. This introduces conventional pay levels to banking.]

Such a clawback would not be subject to averaging or offset one year to the next. And it would be payable in cash.

The idea, Paul explains, is twofold. First, to remove the government from the business of determining what is, or is not, risky—really, the government doesn't know.

Second, to increase the personal risk of failure for senior management, but stopping short of the sword of Damocles of unlimited personal liability. If bankers are venal, why not harness that venality in the public interest?

For the better part of 100 years, and especially in the past five, we have socialized the risks of high finance. All too often, the bankers who take risks don't themselves bear them.

By all means, let the capitalists keep the upside. But let them bear their full share of the downside.

There are problems with both approaches; I would remove shareholders from double liability; they have insufficient knowledge of ongoing activities, but either – just as they stand – would work better and be preferable to government, its regulations, and record of failures – and all financed by the people's money.

These procedures apply only to banks and other financial institutions: farming, small neighborhood restaurants, laundries, service stations, barber shops, beauty salons, local machine shops, mining and construction firms, product manufacturers, to great national, international companies such as General Electric, Boeing Company, etc., would not be involved in any way – unless financial operations are involved.

If government would only look for solutions to business problems from those who live the problems, life in the U.S. would be more stable. Government departed from double liability because it didn't stop bank problems. But the Yale Law School Legal Scholarship Repository has studied bank history from early times to the present, and its conclusions are: Government made a big mistake; it should still be the law.

Banking, Investments, Gambling

In 2011, there were only 7313 financial institutions – as compared to 29,000,000 firms of the other type. That, as I mentioned previously, seems far too few to me.

When banks were allowed to add investment activities and the stock market to conventional banking activities, it was reasonable; investors primarily evaluated stocks for purchase by the quality of the firms, the record of their sales and earnings, their standing in the market, and overall performance. Then they purchased them to hold as investments for extended periods of time. That is no longer true; technology and computers have brought about changes in the financial marketplace.

The market is now being unduly influenced by gamblers armed with huge, extremely fast computers, and they are gambling on short-time, almost instantaneous changes. The stock market is no longer sufficiently stable to operate with conventional banking operations. In addition, banks with the additional financial betting capability, have reached the government's measurement of being too big to fail. A separation of conventional banking from the business of stocks and investments should be entertained.

I began evaluating and buying stock in outstanding firms in 1950. My heaviest had been in the company I worked for, General Electric. I did not have a large portfolio, but it suited me, and I was always reading market news in order to keep up -- and had been reasonably successful. Several years ago, something happened. My stock holdings and new selections began to become questionable. I would read a good report for a company with a good record, check it out, and then watch its value fall. Finally, in late 2007, I decided my procedures were no longer working, and I had better do something else. In 2008 I dumped all my stocks on a financial advisor, and its people replaced all my individual stocks with collections of stocks. It is fortunate that I did; I was heaviest with GE, and the advisor sold it at $33, and it subsequently went down to around $6. Now, instead of betting on the future of individual stocks, we are betting on the future of a collection of firms' stocks and the overall market for all stocks and securities.

And some in the financial marketplace agree. Today, James Freeman's article *On Wall Street, They Call Him an Optimist*, WSJ, Opinion, 7/21/2012, reports on his interview with "the famed guru, [the 79-year old dean of forecasters] from the market glory days." Today, Mr. Wien says that stock trading has replaced investing, with the average holding period for stocks going from eight years in 1960 to seven months today. "I think the public feels that professionals have taken over the market and the playing field isn't level for them," he says.

> … Mr. Wien has identified several causes for the 2008 crisis, including "the dumbest idea in the economic history of the United States." That would be the federal policy that "every American should own their own home."

The Wall Street Journal was the first to introduce me to the high speed trading phenomenon. Since then, articles have proliferated. It is obvious. The stock market is no longer a suitable companion business for conventional banking activities. Perhaps some of its activities are suitable, but I will leave the selection to the bankers.

And then, on the Wall Street Journal Opinion pages, July 25, 2012, Judy Shelton presented an article: *The Soviet Banking System – and Ours*. She wrote:

> Many in America today fear that our nation is going the way of Europe—becoming more socialist and redistributionist as

government grows ever larger. But the most disturbing trend may not be the fiscal enlargement of government through excessive spending, but rather the elevated role of monetary policy.

Our central bank, the Federal Reserve, uses its enormous influence over banking and financial institutions to channel funds back to government instead of directing them toward productive economic activity. For evaluating the damaging effects of this unhealthy symbiosis between banking and government, the more instructive model is the Soviet Union in its final years before economic collapse.

We can draw lessons from the fact that the Soviet Union went bankrupt even as its fiscal budget statements affirmed that government revenues and expenditures were perfectly balanced. Under Soviet accounting practices, the true gap between concurrent revenues generated by the economy and the expenditures needed to sustain the nation was obscured by a phantom "plug" figure that ostensibly reflected the working capital furnished by the Soviet central bank, Gosbank.

The problem for the Soviet government was that financing provided by the state-controlled bank was supporting an increasingly unproductive economy—bailing out unprofitable enterprises that had long since quit producing real economic gains that might have raised living standards. The extension of credit to these entities had little to do with merit or potential usefulness.

The Soviet central bank was making up for the difference between government revenues and government expenditures by creating empty credits to be disbursed by central-planning bureaucrats. By the time Mikhail Gorbachev came to power in 1985, vowing to address the disastrous financial situation of the Soviet Union through "perestroika," or restructuring, the budget deficit being financed through the nation's central bank amounted to more than 30% of total government expenditures.

Lenin had been wise about the uses of banks. Shortly before the October Revolution (The Russian Revolution of 1917 is also called the Bolshevik Revolution or the October Revolution in which the Provisional Government was overthrown by the Bolsheviks), he wrote: "Without big banks, socialism would be impossible. The big banks are the 'state apparatus' which we need to bring about socialism, and which we take ready-made from capitalism."

Those big banks can be easily seen today in America: They're the ones deemed too big to fail because their demise would threaten U.S. financial stability. As mandatory members of the Federal Reserve System, they are vital partners for conducting monetary policy through the purchase and sale of Treasury bonds orchestrated by our central bank, a process known as "federal open market operations." Besides serving as conduits of Fed policy for expanding or contracting the money supply through Treasury debt transactions, commercial banks can also access short-term funding directly from the Fed through its "discount window."

As our own nation's budget deficit has grown substantially larger in recent years—with the shortfall between government receipts and government outlays widening to 34.9% in the enacted budget for fiscal year 2012—our central bank has aggressively stepped up its involvement in financing government spending in excess of revenues.

In 2011, the Fed purchased a stunning 61% of the total net Treasury issuance, thus absorbing a huge portion of the fiscal overhang. Meanwhile, the Fed has been making funds available to member banks at record-low interest rates, targeting zero to 0.25% in the federal funds market and charging less than 1% on primary loans through the discount window.

It's a bad combination: The Fed, a government agency, not only conducts monetary policy through commercial banks using Treasury debt and by extending virtually cost-free lines of credit; it also regulates those same entities. Our nation's depository institutions are at risk of becoming complicit instruments of the federal government rather than private credit-granting companies serving free enterprise.

Washington's dire financial condition is distorting the very nature of banking and defeating the fundamental purpose of financial intermediation. Instead of taking on the risk of making loans to small-business owners, or to individuals wanting to purchase underpriced real estate with future potential, bank portfolio managers have every incentive to play it safe. Why do anything that might raise the eyebrow of the visiting banking examiner?

Even as community bankers feel the subtle pressure to avoid local lending, the distorted incentive structure resulting from the Fed's behemoth presence in banking and finance has its greatest impact among the larger institutions. They can earn more profits by trading sophisticated financial *derivative* instruments and speculating in currency markets rather than engaging in the hard grind of evaluating individual proposals from entrepreneurs seeking investment capital.

(A *derivative* is a contract to buy or sell an asset, or exchange cash based on a specified condition, event, occurrence, or another contract. They are contracts whose value is derived from another asset, which can include stocks, bonds, currencies, interest rates, commodities, and related indexes. The purchasers of derivatives are essentially wagering on the future performance of that asset.)

According to the Bank for International Settlements, more than 75% of the $647 trillion in notional value of outstanding derivatives arises from such contracts linked to interest rates—an indication of the extent to which monetary policy dominates the world of big finance.

(Notional value: If a futures contract obligates a buyer to purchase 250 units of the S&P 500 Index and the index is trading at $1,000, the notional value of the futures contract is $250,000.

Capitalism [our free market], depends on access to capital. It's a sad development that banks have turned away from the noble task of directing financial seed corn to the most promising harvesters of productive endeavor. And that they are drawn instead to playing the Fed's nuanced game of betting on government debt and arbitraging interest-related plays.

Recent suggestions that perhaps the solution is to involve our central bank even more in the lending decisions of banks—by having the Fed grant special funds to American banks for the express purpose of re-lending them to government-approved nonfinancial borrowers—highlight how alarmingly dirigiste [opposed to free-market] the entire system has become. Can central planning be far away?

Ms. Shelton is senior fellow at the Atlas Economic Research Foundation and the author of "Money Meltdown" (Free Press, 1994) and "The Coming Soviet Crash" (Free Press, 1989).

I have never understood the moves Mr. Bernanke and his central bankers were doing with interest rates, but I knew the stimulus stuff was foolish. Now, we are seeing that their other activities have had little success and are highly questionable.

Conventional banking is a business and contributes to the well-being of the business community. If run prudently, business can be good. If I apply for a loan, my past credit record should be reviewed, my income and job security should be looked into, the home I am buying should be evaluated. The interest rate I am paying has to be larger than what the bank is paying, etc. Once upon a time, that was expected. It only changed when government entered the picture with demands, and government-run banks (i.e. Freddie Mac and Fannie Mae) entered the arena.

I have spent an inordinate amount of time investigating and trying to understand derivatives of all kinds. The complexity is amazing, and without extremely high-speed computers, these financial operations could not be possible. And they are as close to gambling as you can get without putting cards on the table. That is sufficient to tell me that our conventional banking system should not be involved. Let those who deal in these areas be separated from banking – completely.

What Is Keeping the Economy Down?

NOT WHAT – who!!!

And the answer is…members of the same group that got us into the mess in the first place.

Our current administration has no understanding of business or the economy. When Obama was elected president, we were in the beginning stage of a recession, and he and all his associates recognized that fact. Yet, instead of selecting people for jobs in his administration with recognized business skills and private-sector knowledge, he thanked those who helped him get elected: he gave them the jobs. Neither he nor any of them had sufficient knowledge of business or the business community. Only actual, close-hand experience in farming, manufacturing, producing something for sale, etc., prepares one with this knowledge.

The President's Cabinet, Constitutionally

The United States president's cabinet was established, constitutionally, to allow president-elects to select some advisors to bring into their administrations; the Founders knew that no single individual could be equipped with the broad range of knowledge required of the chief executive of the Union.

The cabinet now includes the vice president and the heads of fifteen executive departments -- the secretaries of agriculture, commerce, defense, education, energy, health and human services, homeland security, housing and urban development, interior, labor, state, transportation, treasury, and veterans. (Forget Obama's czars; they will disappear when Obama does.)

To face the daunting task of the presidency, each president must select his cabinet carefully; a broad range of experience in business, farming, the military, transportation, medicine, banking, etc., is mandatory. If that experienced advice is lacking, successful solutions to difficult situations that arise become tenuous.

President Obama, with a total lack of business or management experience, did not choose his cabinet and staff wisely. That partially explains his failure to overcome the business recession that began December 1, 2007. America was faced with economic problems when he was elected, and a good leader, absent experience of his own, would have sought members for his staff that had a good range of business experience. We definitely expected him to arrive on the scene equipped with the tools necessary to meet the nation's existing problems. And we knew that he would have little opposition to his proposed solutions -- his party dominated both houses of the Congress.

He arrived the poorest equipped of any president in America's history. Only 8% of Obama's cabinet had had prior private-sector business experience. Before Obama, the lowest level had been that of John F. Kennedy at 30%; the largest, that of Dwight D. Eisenhower, had been 57%. The average of all others had been 44%.

In 2009 Obama could get no help from Congress. Lawyers, teachers, social service and arts people, and professional politicians brought little useful experience to governing. Yet, they made up 81% of the Senate's members; and the House, also poorly equipped at that time, was of little help.

This aggregate of government talent decided to spend trillions of dollars to resolve problems in areas completely foreign to them, and we have now seen the total absence of success. The government now has ownership in banks and businesses, and unemployment, at

9.1% in September, 2011, is 8.2% in May, 2012. Now, questions are beginning to be raised concerning the accuracy of the government's unemployment measurement and reporting procedures.

Nothing the administration or Congress has done or is doing in 2012 will remedy the problem. They keep shipping money out to specific areas to stimulate spending, but that money has to come from the private sector -- they are taking money away from total business-tax income in order to give some to a government-favored sector to spend. And there has been no hope for success – they have too little knowledge of business to choose wisely.

But in 2010 help arrived. The people put Republicans in charge of the House of Representatives, and President Obama could then converse with some people with helpful knowledge. But ignorance knows no bounds. The House members said our problem is too much spending for the income that is possible. Obama and his minions said the problem is not enough income for the spending that is necessary.

And our popular media, in an effort to help, said: Stop this partisan bickering and do something – all the time looking in the wrong direction – obviously waiting for Republicans to stop their obstructionist ways. Obstructionist ways: The House members have been trying to convince the Democrat-controlled Senate to reduce spending. The Democrat-controlled Senate wants to increase taxes on the wealthy, and not because that will help the economy; it's because they think it will help them get reelected. .

We know the current administration has been the problem for the past three years – as it and Congress doled out billions of dollars of our money to favored members of our national community – with nothing in return but smiles from the favored. However, prior to that, we had had a Republican Congress doing some of the same.

The experience necessary for understanding the economy is not generally gained in the financial area; that experience is uniquely different. Those with that background generally know too little about the workings of the businesses that make up the economy. But the knowledge of some in the financial area who have been directly involved with business is valuable. Their ability parallels that of the CEOs of large companies like General Electric, as mentioned previously. They understand business and industry, know profit and loss procedures, and they, with unusual aptitudes for numbers and consulting techniques, have been a good addition to the investment field in the world of business.

Knowledge of the economy is seldom found in universities. But some professors -- leaders, with no useful business experience -- are prepared to teach you all you need to know. Those leaders try to show their advanced students all they need, singly, to replace the activity of all those individual minds that make up our economy. They do not know that that is impossible. A lot of university professors are advising Washington, and the advice is obviously failing – and will usually fail.

Presidents Gerald Ford and Jimmy Carter, from 1973 to 1979, successively, tried to reverse a lasting downturn using the stimulus route, and both failed. President Reagan arrived in 1981, faced with a downturn and inflation, rejected temporary stimulus measures, and proposed permanent income-tax rate reductions and other encouraging stuff. By late 1982 the recession was over; in early 1983 employment and investment began to rise rapidly. Nearly two decades of strong, economic growth ensued.

Compare Obama's preparation for the presidency with that of Ronald Reagan's as he began his presidency. It was recently revealed by George Schultz to the Wall Street Journal and published May 26, 2012 on the Opinion page. It is explained by the Wall Street Journal's introductory Editors' note: *The following are excerpts from a Nov. 16, 1980 memo to President-elect Ronald Reagan from his Coordinating Committee on Economic Policy. Its title: "Economic Strategy for the Reagan Administration."* The memo describes an era similar to our current one in its economic problems and public anxiety, and lays out a strategy to address them.

Sharp change in present economic policy is an absolute necessity. The problems of inflation and slow growth, of falling standards of living and declining productivity, of high government spending but an inadequate flow of funds for defense, of an almost endless litany of economic ills, large and small, are severe, they are not intractable. Having been produced by government policy, they can be redressed by a change in policy...

The complete memo can be read at: http://online.wsj.com/article/SB10001424052970204880404577225870253766212.html

It will be interesting to observe the advice President Obama received when it, too, is revealed in thirty years. But you can be sure the advice was different. And it won't be revealed by persons friendly to Obama; it will be revealed by individuals from the other political side -- individuals intending to injure his reputation, if he has anything of consequence left at that time.

Government can only establish the proper environment for the free market economic system to work, and our founders put us on the right track with our Constitution. We have been straying from it ever since its creation, but we still retain enough of it. Too few of us in the United States realize that our economic system, in addition to making us the richest nation on earth, also substantially improved the lives of all earth's people. And we did it by sharing our economic knowledge and creativity – our business people invited foreign technical people to visit us to learn, our people went to other countries to further their own business interests. Government had little to do with it, other than to stay out of the way and to allow our people to engage.

Our business people, still following their search for the best for themselves and businesses (greed, you know), were responsible for those efforts. In the '60s and '70s we brought European and Asian engineers over here, showed them the advantages of newly developed electronics, automated machines and other systems, and they went back and spread the word.

GE introduced its solid-state replacement for tube-type electronics in September, 1960. Their solid-state digital controls for machine tools and other processes were demonstrated at the International Machine Tool Show in Chicago. And in 1961, Mac McCleary, Numerical Control Sales Manager, Specialty Control Department, Waynesboro, Virginia, informed us (I was product planner, application engineer for the controls) of arrangements made to market the new controls internationally. Our first session to train European and Asian engineers in the new technology was held January 22-26, 1962, followed by February 5-11, February 19-23, March 5-9: and April 3-6. And during this time, of course, other schools were being held for domestic people who had purchased NC machines—for maintenance people, for operators, for programmers, etc. In addition, customer visits for all types of information proliferated—from both user and original equipment manufacturers (machine builders). So it was a very busy time.

And as time passed and markets developed, companies went overseas, participated in foreign markets. They learned much from our way of doing business, and we learned from them how to better market our products in foreign countries. At home, we complained. Many of the jobs that we felt should have been ours began to go overseas. Labor was cheap and the people there needed our business people's products.

Our businesses prospered and overseas communities learned the businesses, and they began to prosper. Recently, in 2012, some of those businesses have begun to come back home. The costs for labor in the European Union and Asia – particularly in China -- have increased. (As the people of those nations get more commercially involved, more of them begin to work, more people are able to purchase stuff, the middle class grows, and competition for labor in the marketplace causes wages to improve.) Regulations of ruler-established governments, still worse than ours, cause delays in activities; transportation costs and delays are a growing problem.

China will not be a problem; its form of government will eventually get in the way. We only need to make our voices heard – necessary to change our government's attitude – and to hold on. We may be coming to a turning point: Fewer companies are leaving for overseas and more coming home.

But one thing is obvious: For the nation to prosper, government has to begin to work with the business world. In the future, if any politician arrives on the scene without expressing that knowledge, we should bid him or her a definite adieu.

When Government Exceeds Constitutional Limits

AS DISCUSSED PREVIOUSLY, essentially all of our nationwide problems with the economy were government created. And in all instances, government was operating outside the limits defined by the Founders. And government was able to do that, constitutionally, because of acts by the U.S. Supreme Court.

Almost all important elements of the Constitution have been violated except the 2Nd Amendment. The National Rifle Association (NRA) through its independent efforts has managed to keep it intact -- in accordance with the Constitution. And the NRA is under assault, comprehensively, by liberal activists to undermine it. Every time a terrible incident involving the use of guns arises, efforts are renewed to make unilateral changes. The fact that the same incidences arise in Europe, sometimes far more terrible and where gun control is heavily in place, is ignored by the liberal elements. They cannot understand: gun control, in Europe or America, does not deter bad guys determined to do their dirty work. We need an NRA in many other areas where Constitutional controls have been, and are being, violated.

How to Recognize a Judge

News reports, daily, refer to members of the Court as "liberal" or "conservative." The liberal members, everyone knows, have been used by politicians to bypass the intent of the Founders and to get us to the present state of undress. We do not need a law degree to understand now that the changes that have occurred would not have been acceptable to the Founders. And if we can recognize that, why don't we begin efforts to remove "liberal judges" from office or to prevent their appointment? The judges get the liberal connotation because their failings have been recognized by our lowest common denominator -- the popular press.

When one of the many four to five decisions has been handed down by our Supreme Court, the press reported, "it was a close decision." That is entirely incorrect. These Justices, all nine, have reviewed the same written materials, have listened to the same verbal observations, pro and con. They are all above average in intelligence and schooling, and the laws involved are as clear to one as it is to all others. Hence, something has to be causing the four – or the five – to render consistently wrong decisions.

It is possible that the Justices, those responsible for the wrong decisions, have a disability that prevents them from processing facts properly. The analytical processes of the four – or the five – could be faulty. We now know that the possession of high intelligence (noted by IQ), advanced degrees from the best schools in the land, and top grades do not guarantee good judgment (smartness). Some people can be in possession of all the advantages, yet their thinking system cannot accept a sequence of detailed bits of related information, process the bits properly, and arrive at a correct answer. We also know that no testing procedures are available that have the ability to measure and predict how accurately an individual under observation can process informative details, analyze them, and arrive at definitive, correct conclusions.

That may be the cause of some Justice errors. Something else, very wrong, is also involved, and the members of one these groups, the four or the five, are breaking their oaths of office. According to Title 28, Chapter I, Part 453 of the United States Code, each Supreme Court Justice takes, and agrees to, the following oath:

> I, [NAME], do solemnly swear (or affirm) that I will administer justice without respect to persons, and do equal right to the poor and to the rich, and that I will faithfully and impartially discharge and perform all the duties incumbent upon me as [TITLE] under the Constitution and laws of the United States. So help me God.

The Justices are supposed to rule in accordance with established, written laws, and much of their work is involved with the Constitution. What about that old manuscript; did the Founders expect it to be revised, as time passed, by a procession of succeeding Justices? No! The Constitution includes definitive procedures on how the people – you and I – could agree to revisions if they became necessary. And they foresaw language changes, such as those for capitalism discussed previously, that would occur during the upcoming years and commented on the need to interpret the words of the Constitution, always, using the original meaning and intent of those who created the words at the time.

David Barton's studies, reflected in his book *Original Intent (The Courts, the Constitution, & Religion)*, WallBuilders Press, 2000) tells us much about the Founders. Noah Webster, responsible for America's first dictionary and the copyright and patent provisions in the Constitution, commented on the language changes that would occur with time, and how misinterpretation and even serious error could result when original meanings are ignored:

> [I]n the lapse of two or three centuries, changes have taken place which in particular passages . . . obscure the sense of the original languages. . . . The effect of these changes is that some words are . . . being now used in a sense different from that which they had [had originally].... [and thus] present

wrong signification or false ideas. Whenever words are understood in a sense different from that which they had when
introduced. . . . mistakes may be very injurious.

To avoid the "injurious mistakes" which may arise from mis-
interpreting the [Constitution], one need simply establish
the original intent of that Amendment. How can this be ac-
complished? As President Thomas Jefferson admonished
Supreme Court Justice William Johnson: "On every question of construction, carry ourselves back to the time when
the Constitution was adopted, recollect the spirit manifested
in the debates, and instead of trying what meaning may be
squeezed out of the text, or invented against it, conform to the
probable one in which it was passed."

James Madison also declared:

I entirely concur in the propriety of resorting to the sense
in which the Constitution was accepted and ratified by the
nation. In that sense **alone** it is the **legitimat**e Constitution.
And if that be not the guide in expounding it, there can be
no security for a consistent and stable, more than for a faith-
ful, exercise of its powers. . . . What a metamorphosis would
be produced in the code of law if all its ancient phraseology
were to be taken in its modern sense.

Justice James Wilson, Justice Joseph Story, and many others of
those early years chimed in with the same words of wisdom, and
documentation that records the Founders' discussions and intent is
easily available to any who need the information.

And Mr. Skousen, *A Miracle That Changed the World*, chimes
in: The government formed by our Founders and documentation to support it was considered a miracle. And it was. It
heralded the rising of a sun. With approximately 5% of the

world's population (in 2006), America has created more new wealth than all the rest of the world combined; we have never suffered from a famine, and have relieved many of those who suffered from it; Americans have been responsible for more discoveries and inventions in science and elsewhere than any nation on earth; America has out-distanced the world in extending the benefits of inventions and discoveries to the vast majority of its people in such fields as medicine, housing, education, power-energy, process automation, transportation, space, aircraft, and agriculture; and we have given more dollars in aid and relief than most of the world nations combined.

Do Liberal Supreme Court Judges Hide Their Infidelity?

But how do Supreme Court Judges, recognized and labeled "liberal" by the entire press – TV, radio, newspapers, Internet commentators and bloggers, magazines -- feel about the Founders' well-known stipulations in this respect. David Souter, during his entire tenure as a Judge made his feelings known, and he had obviously served his entire career as a politician. He was happily engaged in forming a Supreme-Court congress that would be more reasonable than that one defined by the Constitution, whose members we had elected to office.

Mr. Souter, during a Harvard commencement speech, said this:

"...A choice may have to be made, not because language is vague, but because the Constitution embodies the desire of the American people, like most people, to have things both ways. We want order and security, and we want liberty. And we want not only liberty but equality as well. These paired desires of ours can clash, and when they do a court is forced to choose between them, between one constitutional good and another one. The court has to decide which of our approved desires has the better claim, right here, right now, and a court has to do more than read fairly when it makes this kind of choice...."

Souter served on the Court from 1990 to 2009. Factors in his decision to resign in 2009 were due to the election of President Obama, who would appoint a successor in agreement with his unprincipled approach to judging. Souter, appointed by a Republican president, has followed a liberal path, during his entire period of service.

The following examines Souter's approach to judging. To review these and more, Google: *David Souter, Harvard Speech, Rebuttal*

David Souter's Bad Constitutional History
by John O. McGinnis and Michael B. Rappaport
Wall Street Journal, June 14, 2010 Opinion

The former justice's logic, and those that have agreed with him, has been shown, definitively in this case, to be incorrect and has caused infinite damage over the years. Messrs.' McGinnis and Rappaport reported as follows:

At the recent Harvard commencement, retired Supreme Court Justice David Souter attacked what he regards as the "simplistic" model of giving the Constitution a "fair reading." A judge, he said, must determine which of the conflicting constitutional values should become our fundamental law by taking account of new social realities. His remarks were a thinly veiled assault on those who, like Justices Antonin Scalia and Clarence Thomas, think the Constitution should be interpreted according to its original meaning.

Justice Souter actually provided a primer on how not to be a judge. He made up a Constitution that never was to justify a kind of judicial power that was never intended.

One of Justice Souter's two primary examples of the need for justices to avoid simplistic judging (i.e., originalism) is Brown v. Board of Education, the landmark 1954 case barring public school segregation. A central premise of Justice Souter's praise of Brown is that it was dictated not by the Constitution's original meaning but by new social realities.

Thus, Justice Souter seems to excuse the court's reasoning in Plessy v. Ferguson (1898), the case that Brown overruled. Plessy had upheld a law requiring blacks and whites to sit in separate train carriages. He suggests that at the time the law "fairly read" and the facts "objectively viewed" were consistent with the decision.

Wrong. Plessy was a dreadful decision precisely because it neglected the Constitution in favor of giving rein to what the justices saw as their own current social realities.

The 14th Amendment, adopted in 1868, commanded then as it does today that no state shall abridge "the privileges or immunities of citizens of the United States" or deny them "the equal protection of the laws." The Plessy court held that separate but physically equal carriages were reasonable in these circumstances, each race having the equal right to sit with its own race.

The separate-but-equal argument was never persuasive. The 14th Amendment was designed to ensure the constitutionality of the Civil Rights Act of 1866—a law that gave blacks the equal right of contract. Yet the law prevented Homer Plessy from having an equal right to contract for the carriage in which he wanted to sit.

Even assuming there was some ambiguity in the concept of equality, if one had to choose between a description of rights in race-neutral terms (the right of an individual to sit in a particular coach) and race-infused terms (the right to sit with one's own race), the former is the better reading. A core purpose of the 14th Amendment was to eliminate caste legislation, and interpreting rights in racial terms is plainly at odds with that purpose.

At the time of Plessy, however, many leading thinkers thought that current social realities, such as the need to bring Southern states back into the union and a public perception of social differences among the races, required segregation. Thus Plessy provides an excellent example of Justice Souter's preferred method of constitutional interpretation, in which social realities trump the Constitution's original meaning.

A similar failure to follow the Constitution's original meaning also robbed African-Americans of their right to vote. The 15th Amendment (1870) prohibited racial discrimination in voting. But until well into the 20th century the Supreme Court permitted discriminatory laws to be applied, and Congress refused to enforce its plain provisions to assure racial equality at the ballot box.

We should reflect on the great suffering and injustice that a failure to follow the original meaning of the 14th and 15th Amendments caused to so many for almost a century. While we believe that an originalist reading of the Constitution also supports Brown, the salient point here is that Brown would not have had such central importance had the Reconstruction-era amendments been enforced according to their original meaning. The greater economic and voting power that enforcement would have ensured would likely have prevented the caste system of public education in the South.

Justice Souter recognizes that his method of interpreting the Constitution is indeterminate, but he argues that it is necessary to put our trust in justices to reach just results. The historical reality is that this interpretive method permitted justices to create a Constitution of their own contrivance in the service of injustice.

Mr. McGinnis is a professor of law at Northwestern University and Mr. Rappaport is a professor of law at the University of San Diego.

David B. Rivkin Jr. and Lee A. Casey, in USA Today, 6/15/2010, point out: Mr. Souter's actions and others with similar views now have us obeying people -- not laws.

> ...It would be difficult to articulate a decision-making model more antithetical to American democracy and the Constitution's own design. It is often said — by the Supreme Court among others — that we have a "government of laws and not of men." Judges are people, not the living embodiment of the law. When a judge makes the choices Souter suggests, without regard to the Constitution's words and their original meaning, it is the judges who rule and not the law.

> The Constitution's drafters understood this very well and, whatever mistakes they made along the way, they manifestly did not empower the courts to choose freestyle among constitutional values. Their judiciary was to be, as Alexander Hamilton explained at the time, the "weakest" branch of government that could exercise only "judgment," not the awesome congressional power of the purse or the president's control over the military.

> ...This is not to say construing the Constitution is easy; it is not. To the extent there are competing values and ambiguous provisions in our founding document, the Constitution itself prescribes how choices ought to be made. To be sure, as human beings, every judge brings a lifetime of personal experiences, beliefs and prejudices (good and bad) to the task of judging. Wading into the Constitution may well seem like walking through a museum of medieval art, which speaks to us in fundamentally different ways than to our ancestors. But the judge's job, his or her sacred trust, requires disciplining these personal experiences and beliefs toward a faithful

interpretation of the Constitution's text.

Moreover, it is possible to rise above personal preference to fairly interpret that text. No better example can be found than in one of the precedents Souter himself discussed at Harvard, to buttress his core claim that reliance on constitutional text causes bad decisions. In *Plessy v. Ferguson* (1896), the Supreme Court upheld the principle of "separate but equal," establishing the legal basis for generations of racial segregation. But there was a dissent.

The Harlan model

Justice John Marshall Harlan ("the Elder") was a man who passionately believed that the "white race" was superior to all others. Yet, as Justice Clarence Thomas likes to point out, Harlan looked into the Constitution and could not find there, in its words as fairly construed, any basis for separate but equal. The Constitution, Harlan wrote, says the government must guarantee the equal protection of the laws to all. That is what it said, and that is what it meant. Harlan was, of course, vindicated in 1954, when the Supreme Court overruled *Plessy* and rejected the notion of "separate but equal" in *Brown v. Board of Education*.

The bottom line is that bad constitutional decisions, far from being the result of the Constitution's frailty, are caused by the frailties of judges who depart from it....

David B. Rivkin Jr. and Lee A. Casey are partners in Baker & Hostetler LLP and served in the Justice Department under Presidents Ronald Reagan and George H. W. Bush
http://www.usatoday.com/news/opinion/forum/2010-06-16-column16_ST_N.htm

What About Sonia Sotomayor, Recently Appointed?

We now have a nation where a future Supreme Court justice, Sonia Sotomayor, could say publicly in 2001, and still be accepted, "I would hope that a wise Latina woman with the richness of her experiences would more often than not reach a better conclusion [on the bench] than a white male who hasn't lived that life." The woman was saying, in far fewer words, precisely what Souter has said. She should never have been appointed as a justice, because she was saying, "I do not intend to judge as dictated by existing laws and the Constitution, even though I will have sworn to do so."

Souter, Sotomayor, and others of like thinking are missing a very important fact. The Founders did not develop a governing system to satisfy the times and the human condition in which they worked. They had studied the many governments that had been formed and had failed since very early times, and had determined what caused them to fail. None of their failures were caused by a change in the social environment; they had all failed because of the weaknesses of human beings. Some people are born with exceptional greed and the need to rule others.

Souter and Sotomayor swore, with their hands on a Bible, to uphold the Constitution as a judge. He never served as a judge; and she is not serving as one. They obviously aspired to be members of Congress and pursued that course as a Supreme Judge. If they felt some things in the Constitution should be changed, the Founders left us a method to do that. The Constitution can be amended, but the people are allowed a voice – a vote on the changes. But that has proven to be too difficult, so politicians have been choosing a different path. And the selection of wayward judges have made many mistakes much easier.

It is very difficult to evaluate the capability or intent of an individual who aspires to a judgeship. And mistakes are made. Another very recent selection, Roberts as Chief Justice, is now being questioned. Selected by a Republican president, he had appeared to be conservative from his past record – believed in the Constitution and its intent when written – and that is still possible. But every judge selected by Democrat presidents during my lifetime has been analyzed with care: Beginning with Franklin D. Roosevelt, every effort possible has been made to select a liberal – one, such as Souter or Sotomayor, whose objectives all differed from those of the Founders.

Americanthinker.com Adds Information:

> FDR attempted to pack the Court with judges that agreed with his objectives, but the Supreme Court "…functioned in its constitutional role as a guardian against the power lust of the other branches of the federal government. …But since then, the Supreme Court has consistently …allowed federal power to extend into almost every area of public life. … It was not the president or Congress which stripped from state legislatures the power to regulate abortion. It was, instead, the Supreme Court, in an opinion written by one of those notional "Republican" justices which made prenatal infanticide a constitutional right under most circumstances.

> …It was also the Supreme Court which, fifty years ago this June, extended the judicial policy of purging religion from public schools to the point that a nondenominational prayer was verboten. …Although this is sometimes presented as a logical extension of the First Amendment, it was not at all. The First Amendment, of course, speaks only of Congress, not of the states.

Mr. Barton tells us: "Only thirty-nine of the fifty-five delegates signed the Constitution. Termed 'Anti-Federalists' they were worried about too much centralization of power in the federal government and insisted on amendments to protect states' rights. Subsequently George Washington, in his inaugural address, urged Congress to consider how the Constitution should be amended. Congress did so, and the result was twelve proposed amendments specifying exactly what the **federal** government, and **only** the federal government, **could not do**. Of those twelve amendments, ten – the Bill of Rights – were ratified by the States to preserve State autonomy over the issues listed in those amendments."

In 1947, Hugo Black and his Court changed all that. They ignored the fact that the amendments were intended, only, to tell the federal government what it could not do. We should revisit Judge Black with after-life impeachment proceedings -- for him and the four who sided with him in his ruling. They managed to be the first to violate the First-Amendment promises the Founders had made to the State representatives, designed to get them to agree to plans for a Union.

The first ten Amendments to the Constitution only limited the activities of the federal government. The feds could not establish a national religion, but the individual states were free to establish a state religion if the people desired it. Theamericanthinker.com adds to Mr. Barton's observations:

> ...Supreme Court invented out of whole cloth the "Incorporation Doctrine," which provided that every guarantee in the federal Bill of Rights was incorporated into the Due Process Clause of the Fourteenth Amendment. That sounds harmless enough until one recalls that each state already had a state bill of rights which guaranteed freedom of religion. The Supreme Court, quite on its own, was federalizing every personal liberty guarantee which states had already been providing. Moreover, there simply were no cases of real abuse by states of freedom of religion in schools.

> A whole slew of laws before ObamaCare created giant bureaucracies with near-omnipotent powers and not so much as a peep out of the Supreme Court. The EPA, OSHA, and the NRLB (which recently threatened to keep Boeing from relocating from unionized Washington State to right-to-work South Carolina), as well as whole federal departments, like

the Department of Education, which have only the most sur-real connection with Article I powers of Congress, have grown with no real check.

... When is the last time that Ginsberg, Sotomayor, Kagan, or Breyer voted on the conservative side of an important deci-sion? How about "never"? The same was true of Douglass, Black, and the other leftist elements of the Warren Court.

What can we do? ...Congress could also do what Newt Gingrich suggested last year (when so many people thought him nuts) and exercise its regulatory power over the feder-al bench by calling judges and justices to answer for their decisions.

There is more: http://www.americanthinker.com/2012/07/the_supreme_court_is_not_our_friend.html

Now we should begin the process of bidding all of them fare-well. We should begin activities to remove all recognized liberals from judge positions as soon as malfeasance is rec-ognized. The four currently recognized as liberal politicians should be the first to go. The two marginal judges should be watched carefully, and the next evidence of malfeasance should get them on the way out.

Our 236-year record of demonstrated excellence is in danger; our rising sun is now on a setting course. Politicians have successfully changed the meaning of our Constitution substantially; and today – now – we must begin corrective activities.

The misinterpretations of the words of our Constitution over the years have been intentional, as you can see from Souter's words. And unfortunately the judges responsible, lacking the knowledge, intel-ligence, and desire for truth as that possessed by our Founders, have caused the disarray in which we find ourselves today. Politicians, acting as judges, have made the Constitution say what they wanted it to say. And I, and you, by not paying attention, participated in creating the mess.

I, in early years, thought the knowledge of lawyers and judges would be useful in creating laws. But I learned with time, circuitously, that this was not true. To create a good law, it is much more important for a lawmaker to be familiar with a problem and what caused it, than it is to have a background of writing skills. Eventually I realized that lawyers' skills were used, mostly, by politicians and other people to circumvent laws with which they disagreed. We should minimize their numbers in federal and state governments.

Over the years, politicians in Washington, D.C. have selected political candidates for the Supreme Court who, we now know, were unqualified to serve as judges. I have never heard anyone ask a candidate this: When deciding constitutional questions, will you search for the original intent of the Founders? Is the Founders original intent that which you will use in formulating judicial decisions?

You should know that there can never be a conservative court or a liberal court. There is either a court that deals in facts and rules as a court, or you have an additional political institution that creates or changes laws, yet calls itself a court. There can be no conservative judge or a liberal judge – those are political terms. A judge makes decisions based on facts and laws, and when he/ she is labeled liberal, you know you have a politician in judge's clothing.

There are only three judges currently serving on our Supreme Court. They are Justices Alito, Scalia, and Thomas. Four of the remaining six are politicians: Justices Ginsburg, Stevens, Sotomayor, and Breyer -- disgraceful. The remaining two, called sometimes-liberal, are sometimes politicians -- Justice Kennedy and, recently questionable, Chief Justice Roberts.

We must begin to pay attention to the Court's actions and begin to shout when we see something wrong. Our Supreme Court has long been the focus of political effort. Dishonest politicians found it to be far easier to get desired social changes by using equally-dishonest, politically-minded judges than to get them formulated, passed in Congress, and voted in by the people – as specified in the Constitution, the way the Founders intended.

W. CLEON SKOUSEN'S book *A Miracle That Changed the World, The 5000 Year Leap* (National Center for Constitutional Studies, 2009), explains the Founders' thinking. It depicts the rough setup of the Founders' vision of the new republic, shown in the book. The triangles are used to illustrate the distribution of power.

The left triangle shows a National government in total charge (the broad base is the national government). It is at the extreme left and depicts a monarchy (a kingdom, and tyranny). The triangle at the center shows the broad base at the bottom (where the individuals are located) with the people at the commanding position. As the people (Individual) lose power to National (government), the triangle moves to the left and begins to tilt. By the time the people have lost entirely and the government is now a monarchy, the triangle has rotated 180 degrees.

The Peoples' Law is located at the center between the two extremities -- Ruler's Law (Tyranny) at the left, and No Law (Anarchy) on the right). The Rulers' Law shows the National government at the broad base of the triangle -- in complete charge; ruling like a king – top down, no participation from individuals (the people). Peoples' Law shows Individuals (the people) in charge; the triangle's broad base, the ruling base, is now at the bottom – ruling from the bottom up. No Law (Anarchy) shows no triangle at all. No one is in charge.

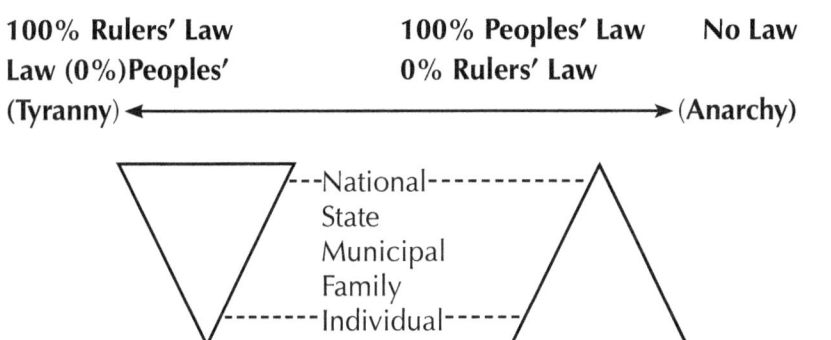

100% Rulers' Law   100% Peoples' Law   No Law
Law (0%)Peoples'   0% Rulers' Law
(Tyranny) ◄————————————————► (Anarchy)

```
---National-----------
   State
   Municipal
   Family
---------Individual------
```

The Founders' Political Spectrum

...Part of the genius of the Founding Fathers was their political spectrum or political frame of reference. It was a yardstick for the measuring of the political power in any particular system of government. They had a much better political yardstick than the one which is generally used today. If the Founders had used the modern yardstick of "Communism on the left" and "Fascism on the right," they never would have found the balanced center which they were seeking....

What Is Left? What Is Right?

It is extremely unfortunate for us citizens that the political philosophy writers and speakers of today have undertaken to measure the various issues in terms of political parties instead of political power. The American Founding Fathers would have considered this modern measuring stick most objectionable.

Mr. Skousen tells us:

...Today, as mentioned, it is popular in the classroom as well as the press to refer to "Communism on the left," and "Fascism on the right." People and parties are often called "Leftist," or "Rightist." We do not really understand what they are talking about.

These political terms actually refer to the manner in which the various parties are seated in the parliaments of Europe. The radical revolutionaries (usually the Communists) occupy the far left and the military dictatorships (such as the Fascists) are on the far right. Other parties are located in between.

Measuring people and issues in terms of political parties has turned out to be philosophically fallacious, if not totally misleading. This is because the platforms or positions of political parties are often superficial and structured on shifting sand. The platform of a political party of one generation can hardly be recognized by the next. Furthermore, Communism and Fascism turned out to be different names for approximately the same thing—the police state. They are not opposite extremes; for all practical purposes, they are virtually identical.... [It is obvious that both are governed from the top down.]

A More Accurate Yardstick

Government is defined in our dictionary as "a system of ruling or controlling," and therefore the American Founders measured political systems in terms of the amount of coercive power or systematic control which a particular system of government exercises over its people. In other words, the yardstick is not political *parties,* it's [the result of both parties' efforts], political *power.*

Using this type of yardstick, the American Founders considered the two extremes to be ANARCHY on the one hand, and TYRANNY on the other. At the one extreme of anarchy there is no government, no law, no systematic control and no governmental power, while at the other extreme there is too much control, too much political oppression, too much government. Or, as the Founders called it, "Tyranny."

In evaluating our parties, we should decide which one desires to take us the farthest to the left – toward a monarchy. We will soon realize that both parties have been taking us there, but the Democrat party is by far the worse culprit. We know that neither is aiming for anarchy.

The objective of the Founders was to discover the "balanced center" between two extremes. They recognized that under the chaotic confusion of anarchy there is "no law," whereas at the other extreme the law is totally dominated by the ruling power and is therefore "Ruler's Law." What they wanted to establish was a system of "People's Law," where the government is kept under the control of the people and political power is maintained at the balanced center with enough government to maintain security, justice, and good order, but not enough government to abuse the people.

Ruler's Law

The Founders seemed anxious that modern man recognize the subversive characteristics of oppressive Ruler's Law which they identified primarily with a tyrannical monarchy. Here are its basic characteristics:

1. Authority under Ruler's Law is nearly always established by force, violence, and conquest.
2. Therefore, all sovereign power is considered to be in the conqueror or his descendants.
3. The people are not equal, but are divided into classes and are all looked upon as "subjects" of the king.
4. The entire country is considered to be the property of the ruler. He speaks of it as his "realm."
5. The thrust of governmental power is from the top down, not from the people upward.

6. The people have no inalienable rights. The "king giveth and the king taketh away."
7. Government is by the whims of men, not by the fixed rule of law which the people need in order to govern their affairs with confidence.
8. The ruler issues edicts which are called "the law." He then interprets the law and enforces it, thus maintaining tyrannical control over the people.
9. Under Ruler's Law, problems are always solved by issuing more edicts or laws, setting up more bureaus, harassing the people with more regulators, and charging the people for these "services" by continually adding to their burden of taxes.
10. Freedom is never looked upon as a viable solution to anything.
11. The long history of Ruler's law is one of blood and terror, both anciently and in modern times. Under it the people are stratified into an aristocracy of the ruler's retinue while the lot of the common people is one of perpetual poverty, excessive taxation, stringent regulations, and a continuous existence of misery.

It is quite obvious that the founders' vision has been lost. The U.S. government is not in total charge – we're not yet approaching a monarchy -- but it has taken over substantial territory from the people. And in all areas, we have seen problems arise that the Founders' knew would occur and even warned us about.

The Founders established a government in which the people were to rule. The people chose the people who were to run the government and make the laws. But they also intended for the people to be involved when any change in the Constitution was considered. And they realized changes in the Constitution would have to be made from time to time, and Article V was included in the Constitution to explain how to make changes.

Whenever two thirds of both Houses deem it necessary, amendments to the Constitution can be proposed. Or, whenever the people, with two thirds of the legislatures of the States, deem it necessary, they can apply for a convention in which amendments can be proposed. Regardless of how an amendment is proposed, however, the final step is acceptance by the people -- ratification. Two methods for ratification are provided. One, Amendments proposed by Congress' and passed by two thirds of the House and Senate, can then be sent to the states for ratification (acceptance). The Congress' proposal must be accepted and passed by three-fourths of the state legislatures (38 states). The other, Amendments proposed by the people (state convention proposal) and passed by two thirds of the state legislatures, can then be voted on in convention, and it must be accepted and passed by three-fourths of the states (38 states).

The difficulty which these procedures entail has been widely discussed, but I have never heard mention of the excellent reasons that they were included. Had we been following the Founders' procedures, instead of Souter's or Sotomayor's, for example, we might have avoided some of the nationwide problems that have beset us from time to time. The Founders were not trying to make the process difficult; they were trying to make sure that any proposed changes in government operations were reviewed by a wide cross-section of the citizenry. The Founders' procedures for changing the Constitution gave the final word to the people. The Founders knew that the populace would have more knowledgeable people in it, by far, than that which could be included in any administration throughout time. They also knew that government would want to make changes due to ignorance or for their own selfish reasons.

This procedure assured that any proposed change in the Constitution could be considered and reviewed by a great cross-section of the nation's inhabitants – university professors, large-business people, small-business people, bankers, financial people, religious people, learned citizens of all types. State legislature people will review with their citizens any proposed federal change that might affect them or their people's livelihood – good or bad. I think it is obvious that the Founders' procedures had a far better possibility of regulating something properly, than trusting in Souter's, the four political judges, and all the other foolish politicians that have had a hand in completely overhauling our government. They got us into the state in which we are currently engaged.

Local Control

Laws originating in Washington, D.C., intended to help local people residing in the cities or towns of states, have caused much grief. "...The Founders always attempted to solve problems on the level where the problem originated. If this was impossible they went no higher than was absolutely necessary to get a remedy." We have seen the problems that arose with the federal government's attempts to help the poor. The federal government cannot supervise the distribution of the funds at the local level, so much of the funds are stolen, misused, or diverted to others not so poor. The federal government was not charged with that responsibility by the Founders – and for good reasons. Until the federal government got involved, neighbors, churches and associated people gave help to the poor, and the number of deserving poor, nationally, was small. The local people knew the poor that needed help, personally, and gave it; they avoided those that just chose not to work, and the latter eventually had to go to work. The federal government is in no position to evaluate local people and their conditions. So, the money it offers the states for the poor is welcomed, is distributed with much less care, and the number of the "needy" has ballooned.

For instance. The Wall Street Journal, in March and June, 2012 articles, noted that the food stamp program, begun in the 1970s, was designed to help the poor -- one of fifty of our citizens. In December, 2011, we were helping one of seven. Obviously we have strayed far beyond helping those who are poor.

In the beginning food stamps were used only as a last resort – shame was felt – and at one time, the recipients had to visit, and frequently join a line, at a local distribution center. That was too public and was discontinued because of the stigma and discomfort that the poor suffered.

Now, the stigma is gone. The Agriculture Dept. runs TV ads alerting people about free food, and the stamps are now being touted in a way to lose weight in a TV campaign. "Food stamps are becoming the latest middle-class entitlement."

The Political Class

The Founders envisioned the best people to serve their country in government were those who had been successful in business, farming, military, etc. And it was to be a service, not a way of life. Benjamin Franklin even proposed that the president be paid no salary, just be paid for his expenses. If the pay had remained minimal, we wouldn't have so many people studying political science. Around 1950, political science began to be taught in universities, and we now have to face it; too many of our government people in Washington have chosen politics as their life's work. They didn't choose politics in order to serve their country, they chose politics as a neat way of making a living.

And it is neat. Think about it. When at home, as Congressional members, they can talk to friendly local business leaders, ask how things are going, and what does the community need? The leaders say things are good, but we could use …. (Fill in the blank.) The member says he can get the community what it could use, but he needs…. (Fill in the blank.) Too many of our government people have arrived in Washington with passable wealth, but were considerably better off when they left.

Religion

David Barton in his book *Original Intent,* explains how the original intentions of the Founding Fathers in writing the Constitution have been changed and how the rules involving religion have been rewritten by Supreme Courts of the past and present – acting in their newer role as political activists. He concentrates on the political-type steps that judges have taken in order to change the Constitution to something they liked better.

> The separation of church and state grew out of a letter that Jefferson wrote to a Baptist congregation after his election as President. "Jefferson committed himself as President to pursuing ... the purpose of the First amendment: not allowing the Episcopalians, Congregationalists, or any other denomination to achieve the [national] establishment of a particular form of Christianity...." In his letter to the Baptists, "...I contemplate with sovereign reverence that act of the whole American people which declared that their legislature should make no law respecting religion or prohibiting the free exercise thereof, thus building a wall of separation between Church and State." He promised the Baptists the government will not interfere with religions' free exercise. He did not say that all religion and state affairs should be separated, as subsequent Courts and political judges have improperly ruled. That is clear. And a reading of his letter, available on the Internet, will show that our current course is improper, and Souter-like directed.

> Another worry at the time was what treatment extreme religious activities such as human sacrifice, polygamy, bigamy, incest, infanticide, etc. would receive from the new government. It was explained that those acts, even if perpetrated in the name of religion, would be stopped by the government; "... since ... they were 'subversive of good order' and were 'overt acts against peace and good order.' The Constitution assured good order."

The changes we have seen in the government's treatment of religions were entirely contrived for political purposes at the behest of atheists and others organized to oppose the beliefs of religious people. The Founders felt, and said this: "The success of the government they had created was dependent upon good, religious people. Without religion, the Founders had little hope for the success of the Republic. John Adams said, 'Our Constitution was made only for a moral and religious people. It is wholly inadequate to the government of any other.'"

We have now seen decades upon decades of the fading of marriage, the birth of more and more children outside of marriage, and a decrease in church membership. While at the same time, we've seen a vast growth of government, as it moves away from the Constitution and its limits and failures. Could we, at this point, acknowledge that it is possible the Founders were correct?

If we want to return religion to the state that was visualized by the Founders, religion will have to find an organization to fight for it with the same zeal as the National Rifle Association has exhibited for the Second Amendment. The NRA has suffered slings and arrows and disappointments of the worst kind, but it has stopped organized gangs from making damaging inroads on the Second Amendment.

Education

Could the failures of education possibly be an underlying cause for the unbelievable growth of the federal government? Some have answered that question in the affirmative.

David Gelernter, WSJ Opinion, 7/2/2012, effectively introduced a message by Mr. Larry P. Arnn. "Almost no one believes that our public schools are doing a passable job of teaching American and Western civilization. ...Many American children have never heard a good word for the United States, the West, Judaism or Christianity their whole lives.

"...Yes, Americanism evolves, and by all means let's change our minds when we ought to. We should always be marching toward the American ideals of freedom, equality and democracy, as we did when we ended slavery, granted women the right to vote, and finally buried Jim Crow. But if we forget our basic ideals or shrug them off, as we are doing today, we no longer deserve to be great. Without our history and culture, we have no identity.

"...Modern American culture is in the hands of intellectuals— unfortunates born with high IQ and low common sense. ...We have failed a whole generation of children."

Mr. Gelernter's article should be read by all: http://online.wsj. com/article/SB10001424052702303734204577468371113471532. html

Larry P. Arnn, President of Hillsdale College, in his book, *Liberty and Learning, The Evolution of American Education,* (2004, Hillsdale College Press) provides some answers. He details what has occurred in education. The Constitution gave the federal government no role in education, so how it did it get involved?

...Andrew Dickson White, founding president of Cornell University, was a speaker at the 100th anniversary of the Constitution of the United States, September 17, 1887. Already the Constitution had survived longer than any other written constitution in human history. The nation living under its dominion had expanded across the plains and the mountains to the Pacific Ocean, and already it was becoming a power in the world. There had to be a grand celebration. The chief one was held in Philadelphia, the city where the nation and its Constitution were born.

The commemoration lasted three days. There were parties, rallies, and parades. There were seminars, speeches, and discussions. At the end there was a gala dinner featuring no fewer than thirteen toasts and responses, some of them lengthy. The opening remarks were given by President Grover Cleveland. Former president Rutherford B. Hayes gave the concluding remarks. Charles Francis Adams, son and grandson of Presidents John and John Quincy Adams, and Philip Sheridan, the swift and deadly Union cavalry commander, were among the speakers. From the founding of the Republic to its salvation through civil war, the great were gathered to recall and honor what had been done. ...Nothing can better portray the importance of education to leading Americans at the midpoint of our history than the inclusion of this speech on this evening's hundred-year event....

White told them how:

Schools now number "hundreds of thousands," teaching "millions on millions," with "hundreds of millions of money lavished upon it by the nation, the municipalities, the rural hamlets, and with a growth of private munificence such as the world has never before seen.

To that point, the nation's people had done that with no direction or help from the federal government. The people knew the value of education for the future of their children and the country; and they had needed no direction from the federal government. They had done this great task with their own efforts and money.

White also wanted to maintain the Constitution in its original form; the nation had obviously done well under the original intent, but he made one little mistake. He approved of a federal Bureau of Education – which would be a single individual – "a servant, not a lord." The individual would merely observe the activities in the different states, record them, and share the good information for the benefit of all.

White had done no study of past governments, how they operated, the influence of human behavior, and how and why past governments had failed – the Founders had made the study. The Bureau was beyond the limits set for federal activities and was unconstitutional. But one little man, no power, only there to do good. What problem could he create? That little mistake of acceptance was the foundation on which the United States Department of Education was built. "Every state now has its own similar department. Below them are districts, also replete with administrators. They compile reports from every school and college in their territory. They keep detailed records at every level. They write detailed standards for every phase and aspect of education. They measure local performance against those standards."

That single individual, in the beginning, saw a state or two doing something well and reported the information to all. But some did not seem to get the message, so a little firmness was necessary, then a law or two, and before the cat blinked twice, we had a house full of servants who had been promoted and had become lords.

And the government, working diligently ever since, watched the costs increase each year as student performance leveled off, and then, still stirring the pot, continued to watch as it began its long deterioration to today.

We now sit at the lowest level of student achievement in the nation's history. And the federal government didn't help us get there; it, with the education unions' help, is responsible for the entire mess.

Congress

The Powers of Congress

We have discussed how badly Congress and the U.S. Government have exceeded their powers, but we haven't shown what the Constitution told them they could do.

The powers of Congress and its law-making realm were enumerated and defined in the Constitution, Article I, Section 8.

> The Congress shall have Power To lay and collect Taxes, Duties, <u>Imposts</u> and <u>Excises</u>, to pay the Debts and provide for the common Defence and general <u>Welfare</u> of the United States; but all Duties, <u>Imposts</u> and <u>Excises</u> shall be uniform throughout the United States;
>
> To borrow money on the credit of the United States;
>
> To regulate Commerce with foreign Nations, and among the several States, and with the Indian Tribes;
>
> To establish an uniform Rule of Naturalization, and uniform Laws on the subject of Bankruptcies throughout the United States;
>
> To coin Money, regulate the Value thereof, and of foreign Coin, and fix the Standard of Weights and Measures;

To provide for the Punishment of counterfeiting the Securities and current Coin of the United States;

To establish Post Offices and Post Roads;

To promote the Progress of Science and useful Arts, by securing for limited Times to Authors and Inventors the exclusive Right to their respective Writings and Discoveries;

To constitute Tribunals inferior to the Supreme Court;

To define and punish Piracies and Felonies committed on the high Seas, and Offenses against the Law of Nations;

To declare War, grant Letters of Marque and Reprisal, and make Rules concerning Captures on Land and Water;

To raise and support Armies, but no Appropriation of Money to that Use shall be for a longer Term than two Years;

To provide and maintain a Navy;

To make Rules for the Government and Regulation of the land and naval Forces;

To provide for calling forth the Militia to execute the Laws of the Union, suppress Insurrections and repel Invasions;

To provide for organizing, arming, and disciplining, the Militia, and for governing such Part of them as may be employed in the Service of the United States, reserving to the States respectively, the Appointment of the Officers, and the Authority of training the Militia according to the discipline prescribed by Congress;

To exercise exclusive Legislation in all Cases whatsoever, over such District (not exceeding ten Miles square) as may, by

Cession of particular States, and the acceptance of Congress, become the Seat of the Government of the United States, and to exercise like Authority over all Places purchased by the Consent of the Legislature of the State in which the Same shall be, for the Erection of Forts, Magazines, Arsenals, dock-Yards, and other needful Buildings;

To make all Laws which shall be necessary and proper for carrying into Execution the foregoing Powers, and all other Powers vested by this Constitution in the Government of the United States, or in any Department or Officer thereof.

Section 9 - Limits on Congress

The Migration or Importation of such Persons as any of the States now existing shall think proper to admit, shall not be prohibited by the Congress prior to the Year one thousand eight hundred and eight, but a tax or duty may be imposed on such Importation, not exceeding ten dollars for each Person.

The privilege of the Writ of Habeas Corpus shall not be suspended, unless when in Cases of Rebellion or Invasion the public Safety may require it.

Habeas corpus (you must present the person in court) is a writ (legal action) which requires a person under arrest to be brought before a judge or into court. This ensures that a prisoner can be released from unlawful detention, in other words, detention lacking sufficient cause or evidence.)

No Bill of Attainder or ex post facto Law shall be passed.

No bill of attainder prohibits all legislative acts, no matter what their form, that apply either to named individuals or to easily ascertainable members of a group in such a way as to inflict punishment on them without a judicial trial. *Ex post facto law* is law which makes criminal an act that was not criminal when done, or which inflicts a greater punishment than the law assigned to the crime when committed.

No capitation, or other direct, Tax shall be laid, unless in Proportion to the Census or Enumeration herein before directed to be taken. (Changed by the 16th Amendment.)

No Tax or Duty shall be laid on Articles exported from any State.

No Preference shall be given by any Regulation of Commerce or Revenue to the Ports of one State over those of another: nor shall Vessels bound to, or from, one State, be obliged to enter, clear, or pay Duties in another.

No Money shall be drawn from the Treasury, but in Consequence of Appropriations made by Law; and a regular Statement and Account of the Receipts and Expenditures of all public Money shall be published from time to time.

No Title of Nobility shall be granted by the United States: And no Person holding any Office of Profit or Trust under them, shall, without the Consent of the Congress, accept of any present, Emolument, Office, or Title, of any kind whatever, from any King, Prince or foreign State.

Amendment 16 - Status of Income Tax, Ratified 2/3/1913. The Congress shall have power to lay and collect taxes on incomes, from whatever source derived, without apportionment among the several States, and without regard to any census or enumeration.

If you review these powers minutely, you will find no mention of Health, Banks, Education, Charity or help for the poor, Religion, or

Press. Religion, which has been under siege for decades is omitted, but the First Amendment enters and defines the federal government's power: It tells the federal government it has no powers -- stay completely out of religion. And the same goes for the Press; it gives Congress no control over radio programs.

Every one of those powers listed in the previous paragraph are today's troubled areas, caused by federal government actions. We will never know, but in all probability many of the nationwide problems suffered in the past would never have occurred had our politicians of all stripes followed the Founders' directions closely during past years.

We are in the midst of a bad downturn, foisted on us by our federal government, which has no idea what the problems are, and could offer no solutions if it did. I think we need to make some changes – some immense changes.

But it has become evident during my lifetime: members of Congress are not going to relinquish the power, prestige, and the good life that their current status gives them. And they make the laws. The peoples' satisfaction with Congress has hovered around 10 to 11% as long as I can remember. If we want changes, the movement has to come from the people. So if we can get a sufficient number of our citizens' complete attention with a proposition, it might be possible to make needed changes.

Congress –Constitutional Limits and Spending

How to stop Congress' excessive spending has become today's outstanding question. Politicians, some of them current Congressional members, have offered a possibility -- vote the bad guys out. Why not; what's wrong with that?

We can't vote them out. The problem isn't with our state's Congressional members; our people are doing a fine job for us and the state. They have brought in millions of dollars from the federal government for worthwhile projects. The problem is with the money the government has given Congressional members of the other forty-nine states; it was spent on foolish, wasteful, projects. We need some way to vote out the misguided Senators and House members of the other forty-nine states.

What about those "millions of dollars of free gifts" that our members of Congress got us from the feds? Were they really free? I hate to burst a bubble, but they were not. The government doesn't have any money (very, very little) to give anybody. Any funds that the fed has, or hopes to have, must come from the private sector (over 92% of it). In order to give us or our state a certain amount of money, the government must take that total amount, plus a little extra, from the only available source – the fifty-states' total private-sector income from taxes. What is the extra for? That's for costs and interest. We sometimes forget, government people don't work for free; we have to pay them. That's why they serve us so well.

What we hadn't considered as we thanked our Congressional member for the gifts was this. Our people weren't the only ones looking for gifts; the Congressional members of every state in the Union – all fifty states – were soliciting and receiving them. We knew that and worried about the other 49 states' wasteful ways, mentioned in the beginning. They were the problem we have seen all along. All of their gifts, in addition to ours, had to be funded by taxes from the private sector. So, as our congressional members appeared at the front door with our gifts from the feds, we were having to send out the back door an equal amount, or more, of our private sector's tax dollars to help pay the total cost of the goodies being distributed to all fifty states.

That would be the complete story if the government's income was sufficient to pay for its spending -- but the income has been insufficient for years. The federal government had to borrow some of the money to give our state its "millions of dollars" of gifts and that given to the other forty-nine, too. That has added to our national debt, and our interest costs increase. And with time and more spending, the debt will grow to a magnitude that is unsustainable, and our children and grandchildren will eventually have to resolve the problem and suffer the pain.

What if some members of Congress do the right thing and refuse to bring gifts of federal money home to their state? They might get some personal recognition, but monetarily they wouldn't help their state much. Those of the other forty-nine states, in proceeding with their efforts to get reelected, will bring home their goodies as usual. The good members' states will get nothing, but must still pay their share of the costs for the goodies for the other forty-nine states.

All those years that the good stuff was coming in the front door, cash to pay for it was going out the back door. We were not profiting as we thought. Your Congressional people surely knew that, but since they didn't tell you, you now know they can't be trusted.

We can't vote for or against those running for election in the other forty-nine states; the only possibility for ending the Congressional careers of long-time members is term limits. Term limits for the Presidency, requiring an amendment to the Constitution, was passed by Congress and ratified by the states. The Congress passed the amendment on March 21, 1947, and it was ratified by the requisite number of states on February 27, 1951. Congress will never act to limit its members' terms of service in the same, conventional, Constitutional way. It would limit their powers and future personal economic possibilities severely. But Article V of the Constitution provides us with another solution.

Article V permits the states to call for a convention for the purpose of proposing Constitutional amendments. Any proposals submitted and accepted during the session, when ratified by three-fourths of the states, become law. Interest is growing. There has never been a Constitutional convention of this type; however, twenty-nine states have now acted to call for one with twenty of them proposing balanced budgets and restraints on the growth of the federal government. Info from www.sweetliberty.org lists them below. We only need 2/3 of the states, 34, to require Congress to begin the process.

States with a Standing-Call for a Constitutional Convention

Alaska	Arizona	Arkansas	Delaware	Colorado
Georgia	Idaho	Indiana	Iowa	Kansas
Maryland	Mississippi	Missouri	Nebraska	Nevada
New Hampshire	New Mexico	North Carolina	North Dakota	Oklahoma
Oregon	Pennsylvania	South Carolina	South Dakota	Tennessee
Texas	Utah	Virginia	Wyoming	

Obama and the Democrats in Congress have told us on numerous occasions the rich aren't paying their fair share of taxes. But he hasn't shown you that. Ari Fleischer tells us the true story in his July 22, 2012, Wall Street Journal article, *The Latest News on Tax Fairness*

A new Congressional Budget Office report shows the share of taxes paid by the top 20% has gone up over the last thirty years, while the share of taxes paid by everyone else has gone down.

If fairness in paying taxes means the amount you pay is based on the amount you earn, then the only group in America paying at least a "fair share" is the top 20%—people who make more than $74,000. For everyone else, the tax code is a bargain.

You wouldn't know this from President Obama's rhetoric, but our tax system, according to a recent report by the Congressional Budget Office (CBO), is incredibly progressive. Consider: The top 1% of income earners pay an average federal tax rate of 28.9%. (See the nearby table.) The average federal tax rate on the top 20% is 23.2%. The 20% of taxpayers earning between $50,100 and $73,999 pay an average 15.1%, and so on down the line. The CBO report includes payroll as well as income taxes paid.

There's also another way of looking at fairness, and that's the tax burden. Here, consider the top 20% of income earners (over $74,000). They make 50% of the nation's income but pay nearly 70% of all federal taxes.

What's So Fair About This?

No matter how you slice the data, the rich pay a disproportionate share of federal taxes.

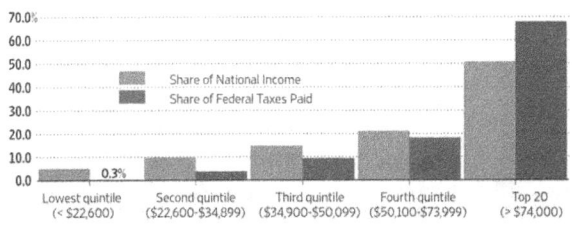

Quintile	Average Tax Rate
Top 1%	28.9%
Top 5%	24.1
Top 20%	23.2
Fourth Quintile	15.1
Third Quintile	11.1
Second Quintile	6.8
Bottom 20%	1.0

Source: Congressional Budget Office, "The Distribution of Household Income and Federal Taxes, 2008 and 2009," July 10, 2012

The remaining 30% of the tax burden is borne by 80% of the taxpayers, those who make less than $74,000. In short, this group's share of taxes paid, 30%, is lower than the share of income they earn, 50%.

Yet President Obama says that "for some time now, when compared to the middle class," the wealthy "haven't been asked to do their fair share."

He's right that the system isn't fair, but not because the top 1% pay too little. It is because they pay too much.

Mr. Obama has said that some wealthy employers pay a lower tax rate than their secretaries. True, some are able to lower their effective federal tax rate by giving millions to charity. Or because they derive much of their income as capital gains or from tax-free municipal bonds.

But middle- and low-income Americans who do not invest also pay lower rates thanks to the deductions they receive, such as a $1,000 per child tax credit (which phases out for couples who make more than $110,000), or the Earned Income Tax Credit, which no one making more than $50,000 is supposed to receive.

The CBO report ("The Distribution of Household Income and Federal Taxes, 2008 and 2009") covers the years 1979-2009. It makes plain that the impression conveyed by the president about what upper-income Americans pay in taxes does not hold up to scrutiny.

First of all, the share of taxes paid by the top 20% has gone up over the last thirty years, while the share of taxes paid by everyone else has gone down. It has gone up despite the tax cuts enacted by President Clinton in 1997 and by President Bush in 2001 and 2003. But that makes no difference to the president. The only group of taxpayers he calls on to "sacrifice" are those already doing all the tax sacrificing.

The top 20% in 1979 made 44.9% of the nation's income and paid 55.3% of all federal taxes. Thirty years later, the top 20% made 50.8% of the nation's income and their share of federal taxes paid had jumped to 67.9%.

And the top 1%? In 1979, this group earned 8.9% of the nation's income and paid 14.2% of all federal taxes. In 2009, they earned 13.4% of the nation's income but their share of the federal tax burden rose to 22.3%.

Meanwhile, the federal tax burden on middle- and lower-income earners is lighter. In 1979, the bottom 20% paid barely any taxes at all, just 2.1%. Now their share of taxes is a minuscule 0.3%. The burden on the middle-income earners ($34,900 to $50,100) has dropped too. In 1979, they paid 13.6% of all federal taxes; in 2009 they paid 9.4%.

One reason our country is so divided is because the president keeps dividing us. If taxes need to be raised to fight a war or fund a cause, the president should ask everyone to pitch in. If the need is national, the solution should be national—and that includes all of us.

But that's not how Mr. Obama governs. We learned during the 2008 campaign that he believes in spreading the wealth around. And recently we learned he doesn't believe that successful people made it on their own. Without the government, the president tells us, job creators and entrepreneurs would not be able to make it in America.

It's really the other way around. Without job creators and the successful, the government wouldn't have any money. So next time Mr. Obama meets someone in the top 1% or even the top 20%, instead of saying they're not paying their fair share, he should simply say thank you.

Mr. Fleischer, a former press secretary for President George W. Bush, is president of Ari Fleischer Communications.

Term Limits for Congress – Both Houses

Raise your voice in your state; get something going. Term limits for all Congressional members should be one of the proposed amendments. Most of the problems we face today were caused by misguided actions of members, both branches of Congress, who stayed too long.

The Democrats ruled Congress for over forty years, before the Republicans finally took over in 1994. The Republicans made some much-needed changes. By 2006, however, the Republicans had demonstrated the same weaknesses as those of the Democrats. It took almost twelve years for Republicans to show that they are just as susceptible to the corruptive influence of power as the Democrats had been before them. And this history clearly demonstrated that we need to make some changes.

We have too many professional politicians – Democrats, Republicans, Independents, and other parties. Most of them, early on in life, had politics in mind, tailored their education to that end, and have only political experience in their background of capability. This limited breadth of experience impedes their ability to properly analyze and resolve problems that confront the nation. Even worse, they fear to consider recognized major problems such as Social Security and immigration because a solution might jeopardize their lifetime careers. A large number of this type of politician has inhabited the House and Senate for decades. Their skills, limited to politics, consist of the ability to talk persuasively, raise money from those that have it, bring government larder home to their State, and to get re-elected.

Our founding fathers visualized Congress as being comprised of talented citizens with experience in industry, engineering, law, agriculture, economics, etc., who, after serving successfully in their fields, would choose to offer their knowledge and skills in service to their government. They could see that the great supply of these citizens and the broad range of individual skills, if utilized in government, would assure a successful future for our Republic.

They foresaw that Congress would enhance the importance of Congressional seats to today's height, but I doubt that they foresaw the development of professional politicians -- individuals that view politics and a congressional seat as a satisfying total career. People who envisioned politics as a way to make their living, got a boost beginning in about 1950. Colleges and universities started teaching political science then.

The time has come for us to change the makeup of the House and Senate, and this will require Amendments to the Constitution. Congress, who makes the laws, limited the President's service to two terms, and the law was ratified by the states in 1951. Congress, whose members make the laws, enjoy their status and have demonstrated no desire to make changes. We will have to help them along.

Term limits for members of Congress have been proposed, and twelve years has been suggested as the limit. That seems to be a good number; some have noted that it took almost twelve years for Republicans to show their political stripes.

I propose two terms for the Senate (twelve years) and four terms for the House (eight years).

Let's start a movement.

Tom Shipley
Birmingham, Michigan